T0094180

.

Power Systems

For further volumes:
http://www.springer.com/series/4622

Paweł Szcześniak

Three-Phase AC–AC Power Converters Based on Matrix Converter Topology

Matrix-Reactance Frequency Converters Concept

 Springer

Paweł Szcześniak
Institute of Electrical Engineering
University of Zielona Góra
Zielona Góra
Poland

ISSN 1612-1287   ISSN 1860-4676   (electronic)
ISBN 978-1-4471-4895-1   ISBN 978-1-4471-4896-8   (eBook)
DOI 10.1007/978-1-4471-4896-8
Springer London Heidelberg New York Dordrecht

Library of Congress Control Number: 2012953565

Printed on acid-free paper

Springer is part of Springer Science+Business Media (www.springer.com)

Preface

The aim of this monograph is to present a concise conception of a new family of modern power frequency converters called matrix-reactance frequency converters. Currently used direct frequency converters, without DC electrical energy storage elements, have some disadvantages, such as the voltage transfer ratio limit of 0.866 in the majority of topologies and control strategies. Because of this these converters cannot be universally used in the industry. For example, in the case of the variable speed drive system for induction motors, a reduction in the supply voltage by 10 % means 20 % loss of torque capability, which is unacceptable in most applications. However, the advantage of this kind of converter is the elimination of a large and expensive DC energy storage. The converters presented in this book also do not have DC energy storage. What is more they enable the buck-boost voltage transformation. The topologies of the presented matrix-reactance frequency converters are based on the three-phase unipolar buck-boost matrix-reactance chopper with source or load switches arranged as in a matrix converter. This approach gives the possibility to obtain an output voltage greater than the input one (similarly as in a matrix-reactance chopper) and a frequency conversion (similarly as in a matrix converter). Nine new topologies of matrix-reactance frequency converters based on boost, buck-boost, Ćuk, Zeta or SEPIC structures are presented.

This monograph is composed of seven chapters, the organization of which can be understood as follows: the first chapter presents a short introduction on the state of the art and future trends in power frequency converters. The actual design tendencies in modern power electronic converters are also discussed here.

The second chapter presents a review of the most important AC–AC frequency converters without DC electrical energy storage as well as basic topologies of hybrid and with DC energy storage element converters. In this chapter the topologies, general operation and properties of this kind of converter are discussed. Mainly attention is paid to direct matrix converter topologies (voltage and current source matrix converters, multilevel matrix converters) and indirect matrix converters (sparse, very sparse and ultra sparse). A large part of the chapter is dedicated to explaining the design and function of a matrix converter. A matrix converter consists of nine bi-directional switches as the main power elements, and creates a variable load voltages with setting frequency. It is also referred as an "all-silicon solution", because it does not have any large energy storage elements. The state of the art in matrix converter technology will be presented in this part of the text, with particular emphasis on control techniques, commutation methods and practical circuit realisation. This part also discusses a conception of converters based on matrix-reactance choppers, which are later called matrix-reactance frequency converters. A separate subchapter is devoted to hybrid frequency converters with small-sized DC electrical energy storage elements.

Chapter 3 describes the concept of a new family of matrix-reactance frequency converter topologies. In addition, a general description of such converters is also presented. The control techniques for these converters, based on the low frequency Venturini method, in particular, are exposed. Moreover, another concept of control strategies are suggested.

In Chap. 4 the averaged state space models of the discussed matrix-reactance frequency converters are described. It should be noted that the models, as a result of averaging, are continuously non-stationary ones, because the average value switch state function of matrix switches are time-varying. In order to obtain a stationary averaged state space model, a two-frequency form (dq) transformation is used. The aim of this chapter is to show stationary mathematical models of all matrix-reactance frequency converters. Furthermore, one part of the chapter includes the solution to the stationary averaged equations, based on stationary averaged models, in steady and transient states. Based on this solution the steady and transient state time waveforms of the averaged state variables are described, and steady-state characteristics are drawn.

Chapters 5 and 6 present some analytical, simulation and experimental test results. Analytical, test results are obtained from the solution presented in the previous chapter. Steady and transient state average time waveforms of currents and voltages in matrix-reactance frequency converters are shown. Furthermore the static characteristics are also presented. Additionally, a simulation verification of the first of two matrix-reactance frequency converters with buck-boost topology has been carried out with the use of a drive system with an inductor cage motor. A simulation study was carried out using the PSpice program. Furthermore, a 1 kVA matrix-reactance frequency converter with buck-boost topologies (two topologies) has been constructed to verify experimentally the concept. The obtained experimental and simulation test results confirm the theoretical analysis. The experimental test results are also shown in this section. Finally, in Chap. 7 conclusions and other comments are made.

Zielona Góra, August 2012   Paweł Szcześniak

Acknowledgments

The author thanks the Institute of Electrical Engineering members for their help and technical advice on this monograph.

Special thanks go to Professor Zbigniew Fedyczak.

Contents

1 Introduction 1
   References 8

2 Review of AC–AC Frequency Converters 17
   2.1 Introduction 17
   2.2 Frequency Converters with a DC Energy Storage Element 19
   2.3 Frequency Converters Without DC Energy Storage Element 22
       2.3.1 Introduction 22
       2.3.2 Direct AC–AC Frequency Converters:
             Matrix Converter 23
       2.3.3 Indirect AC–AC Frequency Converters Without
             DC Energy Storage Elements 65
       2.3.4 AC–AC Frequency Converters Based
             on Matrix-Reactance Chopper Topologies 72
   2.4 Hybrid AC–AC Frequency Converters 74
   2.5 Summary of Topology Review 77
   References 79

3 Concept of Matrix-Reactance Frequency Converters 87
   3.1 Introduction 87
   3.2 Topology Generation 88
   3.3 Topologies of Matrix-Reactance Frequency Converters
       with Voltage Source Matrix Converter 95
   3.4 Topologies of Matrix-Reactance Frequency Converters
       with a Current Source Matrix Converter 98
   3.5 Control Strategies 100
   3.6 Chapter Summary 103
   References 103

4   Modeling of Matrix-Reactance Frequency Converters   107
    4.1   Introduction   107
    4.2   Averaged State-Space Model   108
    4.3   Stationary State-Space Averaged Model: dq Transformation   110
    4.4   Solution of Stationary State-Space Averaged Equations   112
    4.5   Mathematical Models of Matrix Reactance
         Frequency Converters   113
    4.6   Chapter Summary   124
         References   124

5   Property Analysis   127
    5.1   Introduction   127
    5.2   Steady-State Analysis   128
    5.3   Transient State Analysis   137
    5.4   Drive System Application   140
    5.5   Chapter Summary   146
    References   149

6   Experimental Investigation   151
    6.1   Introduction   151
    6.2   Practical Implementation   151
    6.3   Experimental Results   159
    6.4   Chapter Summary   163
    References   167

7   Summary of Book   169
    References   171

Index   173

Acronyms

3D	Three-Dimensional Space
AC	Alternating Current
A/D	Analog to Digital Converter
ARCP	Auxiliary Resonant Commutated Pole
B2B	Back-to-Back
CSI	Current Source Inverter
CSMC	Current Source Matrix Converter
CSR	Current Source Rectifier
D/A	Digital to Analog Converter
DC	Direct Current
DPFC	Direct Power Frequency Converters
DSP	Digital Signal Processor
dq	Direct–Quadrature Transformation
EMC	Electromagnetic Compatibility
EMI	Electromagnetic Interference
FACTS	Flexible AC Transmission System
FCC	Forced Commutated Cycloconverters
FPGA	Field Programmable Gate Array
GTO	Gate Turn-Off Thyristor
IGBT	Insulated Gate Bipolar Transistor
ILMC	Inverting Link Matrix Converter
IMC	Indirect Matrix Converter
I/O	Input/Output
JFET	Junction Gate Field-Effect Transistor
JTAG	Joint Test Action Group
LC	Resonant circuit consists of an inductor (L) and a capacitor (C)
LPF	Low-Pass Filter
LSCS	Load Synchronous-Connected Switches Set
MC	Matrix Converter
MCS	Matrix-Connected Switches Set
MIMC	Multilevel Indirect Matrix Converter

MLMC	Multi Level Matrix Converter
MMC	Modular Matrix Converter
MOS	Metal-Oxide-Semiconductor
MRC	Matrix-Reactance Chopper
MRFC	Matrix-Reactance Frequency Converter
MRFC-b	Topology of MRFC Based on Boost MRC
MRFC-I-b-b	First Topology of MRFC Based on Buck-Boost MRC
MRFC-II-b-b	Second Topology of MRFC Based on Buck-Boost MRC
MRFC-I-c	First Topology of MRFC Based Ćuk MRC
MRFC-II-c	Second Topology of MRFC Based on Ćuk MRC
MRFC-I-z	First Topology of MRFC Based on Zeta MRC
MRFC-II-z	Second Topology of MRFC Based on Zeta MRC
MRFC-I-s	First Topology of MRFC Based on SEPIC MRC
MRFC-II-s	Second Topology of MRFC Based on SEPIC MRC
MTO	MOS Turn-Off Thyristor
NCC	Naturally Commutated Cycloconverters
PC	Personal Computer
PCI	Peripheral Component Interconnect
Ph.D.	Doctor of Philosophy—for the Latin *Philosophiae Doctor*
PWM	Pulse-Width Modulation
RAM	Random Access Memory
RB-IGBT	Reverse Blocking Insulated Gate Bipolar Transistor
RLC	Resonant circuit consists of a resistor (R) an inductor (L) and a capacitor (C)
SCS	Synchronous-Connected Switches set
SEPIC	Single-Ended Primary-Inductance Converter
SMC	Sparse Matrix Converter
SPFC	Static Power Frequency Converters
SSCS	Source Synchronous-Connected Switches Set
SVM	Space Vector Modulation
VSI	Voltage Source Inverter
VSMC	Very Sparse Matrix Converter
VSMC	Voltage Source Matrix Converter
VSR	Voltage Source Rectifier
UPS	Uninterruptible Power Supplies
USB	Universal Serial Bus
USMC	Ultra Sparse Matrix Converter
ZCS	Zero Current Switching
$\alpha\beta\gamma$	Stationary Orthogonal Coordinate Systems

Chapter 1
Introduction

Solid-state AC–AC converters are widely used in a number of applications such as adjustable speed drives [16, 17, 61], AC–AC transmission [13, 50, 105], uninterruptible power supplies (UPS) [12, 63], aircraft converter systems [10, 28] and renewable energy conversion systems [1, 11, 19, 27, 62, 109]. An important area of AC–AC conversion concerns variable speed drive systems for induction motors which currently account for about 50 % of electricity consumption [16]. Generally, the frequency converter (frequency converters) converts AC electrical power of one frequency into AC electrical power of another frequency. Typically, such units are used to convert between 50/60 Hz source frequency into 0–800 Hz on the load side [61]. The term frequency converters does not explain fully the functional capabilities of this kind of power converter. In addition to the possibility of load voltage frequency control relative to the source voltage frequency, these kinds of converters also have the capability to control amplitude of load voltages, to control the displacement angle of the load voltages relative to source voltages, to control the displacement angle between source currents and voltages (input power factor) and the bidirectional (or only unidirectional) power flow control through the converters [47].

The most desirable features in frequency converters are the possibility of generating load voltages with arbitrary amplitude and frequency (also with higher amplitude than the amplitude of source voltages), sinusoidal source and load currents and voltage waveforms, the possibility to provide unity power factor for any load, and finally, a simple and compact power circuit [120]. Over the past decades a number of different PWM AC–AC frequency converter topologies have appeared in the literature [2, 3, 6–9, 14, 31, 43, 44, 47, 52, 56, 58, 64–67, 69–74, 78, 80, 82–85, 89, 92, 93, 101–104, 107, 112, 114, 115, 120–122, 124–126, 131, 132]. AC–AC converters are commonly classified as either an indirect converter which utilizes a DC link between the source and load or a direct converter that provides direct conversion [14, 72–74, 107].

Converter systems with either a voltage or current DC link are commonly used in industrial applications. Such converters are known as indirect frequency converters with a DC energy storage element and have been investigated extensively for many years [52, 73]. The indirect frequency converter with a DC energy storage element

P. Szcześniak, *Three-Phase AC–AC Power Converters Based on Matrix Converter Topology*, Power Systems, DOI: 10.1007/978-1-4471-4896-8_1, © Springer-Verlag London 2013

and pulse-width modulation (PWM) was first reported in the mid-1970s [92, 93] and concerned thyristor inverters. Further, the frequency converter with a transistor inverter was developed. The DC link has two implementation forms: a voltage DC link with a capacitor C_{DC} as DC electric energy storage element (with voltage source inverter VSI) and current DC-link with an inductor L as DC electric energy storage element (with current source inverter CSI) [52]. In the case of the voltage DC link, in the simplest unidirectional case, it is implemented by a diode bridge on the rectifier side. An AC–AC converter with bidirectional power flow can be implemented by coupling a PWM rectifier and a PWM inverter to the DC link. In the case of the current DC link the current source inverter would need switches with reverse blocking characteristics. A current source inverter is implemented as a series connection with an IGBT and a diode. Due to the DC link storage element, there is the advantage that both converter stages, rectification and inverter, are to a large extent decoupled for control process. On the other hand, the DC link energy storage element has a relatively large physical volume. Furthermore, in case of the VSI, the application of electrolytic capacitors is a major cause of reduced converter lifetime.

Improvements in power semi-conductor switches over the last few years have resulted in the development of many AC–AC converter devices without DC electric energy storage elements [2, 7–9, 43, 44, 47, 56, 58, 71–74, 78–85, 89, 101–104, 107, 112, 114, 115, 120, 121, 124, 125, 131, 132], which are referred to as direct power frequency converters (DPFC) or static power frequency converters (SPFC). DPFC basically consist of an array of static power switches connected between the source and load terminals. The basic operating principle is synthesis of the load voltage waveforms and source current waveforms from selected segments of the input voltage waveforms and output current waveforms, respectively. The first works on direct power frequency converters connecting the converters with thyristor switches, which are known as naturally commutated cycloconverters (NCC) [47]. Then, the real development of DPFC starts with the introduction of power transistors for implementing the converter switches, and these converters are named forced commutated cycloconverters (FCC).

There are many topologies of direct power frequency converters. The most common is the matrix converter (MC). The first study of MCs was presented in 1980 by Venturini and Alesina [112]. The authors presented the basic configuration of a power circuit as a matrix of bidirectional power switches that connect each load phase to each source phase, and they introduced the name matrix converter and presented the first control theory. For good performance the matrix converter should have an input filter, which minimises the high frequency components in the input currents and reduces the impact of the perturbations of the input grid. The control theory of MCs as proposed in [112] is known as Venturini modulation. This modulation is based on a low frequency modulation matrix which describes the low-frequency behaviour of the MC voltages and currents, and has an output voltage limitation equal to half of the input voltage ($q = 0.5$) and restricted input power factor control. Alesina and Venturini in works [4, 5] presented a method to increase the maximum voltage transfer ratio to 0.866 through the inclusion of third harmonics in the setting input current and output voltage waveforms. This modulation is known as the optimum

Venturini method or improved Venturini. In both Venturini modulation methods, also known as the direct transfer function approach, the output voltages are obtained by the multiplication of the modulation matrix using the input voltages.

The voltage transfer ratio less than one is the main disadvantage of the matrix converter [97, 120]. Over the past few years, the problem of low voltage transfer ratio has stimulated laborious research focused on the development of new control strategies which increase the amplitude of load voltages. In 1983, Rodriguez proposed in [96] a new control technique which is based on an approach with a fictitious DC bus. In this modulation the input voltages are first rectified to a fictitious DC bus and then are inverted to the output voltages with the required frequency. In this method, in the inverter stage the switching is arranged so that each output line is switched between the most positive and most negative input lines using a pulse width modulation (PWM) technique, similar to that in voltage-source inverters. This control strategy is developed by Ziogas et al. and is discussed in detail in [133, 134]. A maximum voltage transfer ratio up to 1.053 is possible to obtain in a matrix converter using this method. Unfortunately, the higher voltage transfer ratio is achieved with low frequency distortion in the source current and load voltages. This concept is known as the indirect transfer function method.

In 1987, Roy and April proposed in [100] the new "scalar" control strategy, where the switch actuation signals are calculated directly from measurements of the phase input voltages. Similar as in Venturini methods the voltage transfer ratio in linear conditions is limited to 0.5 and its increase to 0.866 is related with low frequency distortion in source and load variables [98, 99]. Furthermore, these scalar methods have limitations in input power factor control. A similar approach is used by Ishiguro. The switch actuation signals are calculated directly from measurements of the line-to-line input voltages. The first approach, where the three line-to-line source voltages are taken into account for calculation of control signals, was presented in [57]. The maximum voltage transfer ratio of the proposed algorithm is limited to 0.75. In the second approach only two line-to-line voltages are switching. By introducing the two-phase switching method, reduction in the switching frequency and increase in the voltage transfer ratio to 0.866 are obtained. The input power factor is also controlled with limitations. A comprehensive treatment of both voltage transfer ratio and input power factor aspects of the scalar method is contained in work [91]. The maximum voltage transfer ratio is also equal to 0.866, but with a wider range of input power factor control.

The space vector modulation (SVM) approach was first exploited by Huber et al. in a series of papers [53, 54] in which the principles of space-vector modulation were applied to a matrix converter. The SVM approach is based on the instantaneous space-vector representation of input and output voltages and currents. The SVM approach has been successively developed in matrix control with direct and indirect approaches [15, 20–25, 48, 49, 68, 87, 88, 95, 97, 116–120] and is the most popular control of the matrix converter. The main fields of research concern the control of the input power factor regardless of the load [20, 21, 87], to reduce the number of switch commutations in each cycle period [20, 25, 48, 49, 88, 118, 119], application in drive systems [68, 81, 90, 95, 116, 117, 123] and unbalanced or nonsinusoidal supply voltage conditions [15, 20, 22, 23]. Generally, the SVM algorithm for matrix converters has the capability to achieve full control of both the output voltages and the instantaneous input current displacement angle (input power factor). The maximum output voltage is equal to 0.866 times the source voltage.

Noteworthy is the control strategy for the matrix converter presented in [25], named the duty-cycle space vector approach presented by Casadei et al. In this paper a very efficient mathematical approach for the analysis of matrix converter modulation techniques can be developed by using the space-vector notation and introducing the concept of "duty-cycle space vector." Using this approach, there are three degrees of freedom available, viz., defining the modulation law, allowing the control of the instantaneous values of the output voltages and input power factor to be obtained [3]. The maximum voltage transfer ratio is equal to 1.155. This level of output voltage is obtained by the high rate of switching during the sequence time T_{Seq}. Consequently, the switching losses are increased.

Apart from those listed above from the most widely known control strategies of matrix converters, in the technical literature there are a few less well-known ones, e.g.: sliding mode control [94], carrier-based PWM control [127] and in recent times the very popular predictive control [110, 111].

It should be noted that in the presented control strategies (except those presented in [25, 133, 134]) the maximum voltage transfer ratio is equal to 0.866. In control strategies presented in papers [25, 133, 134] the voltage transfer ratio is greater than one, but the output voltages and source current waveforms have low frequency distortion.

Another major problem in matrix converters is the simultaneous commutation of controlled bidirectional switches. Commutation is very difficult to achieve without generating overcurrent or overvoltage spikes that can destroy the power semiconductors. Fortunately, this problem was solved with the development of several multistep commutation strategies that allow safe operation of the switches [18, 26, 29, 30, 51, 77, 86, 106, 119, 120, 128–130]. The commutation strategies proposed in the literature have varying numbers of steps (one [129], two [29, 30, 128, 130], three [26], four steps [18]) and varying degrees of knowledge about switch signals (current, voltage or current and voltage).

A matrix of nine bidirectional power switches can be arranged for voltage source or current source on the input side. The conventional matrix converter proposed by Venturini and Alesina in the work [112] is named voltage source matrix converter (VSMC). Whereas, the current source matrix converter (CSMC) can be realised by a matrix of bidirectional power switches connected to the current source [78, 79]. The practical realisation of CSMC has inductors on the input side and capacitors at the output side [78]. The function of the switches (transfer matrix) in CSMC are transposed, compared to VSMC, because the switches are fed by current sources and, for this reason, the source inductances must never be opened. On the other hand, the load has a capacitive nature and, for this reason, the output terminals (capacitors) should not be closed. These rules are the opposite of the rules relating to VSMC. The principles of CSMC cannot be found analysed in detail in technical papers, though in the works [46] and [78] the basic operation and principles are discussed. As is presented in [46] and [78] the voltage transfer ratio in CSMC is greater than one. It is a very attractive advantage, but due to the current sources, the CSMC topology cannot be developed. In this converter, theoretically, all the control strategies that are used in VSMC [120] may be used. In the cited references the classical Vanturini [78] and space vector modulation [46] are introduced.

Alternative structures to voltage source matrix converters have been investigated in [56]. The proposed topology is known as indirect matrix converter (IMC) or two-stage matrix converter and it maintains the same input–output performances as the VSMC. An IMC is a hardware implementation of an MC with indirect modulation, as proposed in [60]. An indirect matrix converter consists of a three-phase to two-phase matrix converter as the current source rectifier (CSR) and conventional voltage source inverter. The DC bus between CSR and VSI do not include DC electrical energy storage devices. Because the IMC does not have any DC energy storage element, for the synchronisation of pulses for CSR and the VSI it is therefore highly important to maintain power balance with sinusoidal supply currents [65]. Most research on IMC has been done using triangular wave voltage command modulation [55] or space vector modulation [59, 60]. The IMC consists of separated source and load stages and offers the same benefits and disadvantages as the direct MC. The voltage transfer ratio is also equal to 0.866.

The IMC also provides an option to reduce the number of switches of the source rectifier bridge. The several topologies of IMC, with a reduced number of power switches, has been presented by Kolar et al. in papers [72]. Because the IMC four-quadrant switch current source rectifier could operate with both DC link voltage polarities, there is a possibility to reduce the rectifier stage circuit complexity. The first topology of IMC with reduced number of power switches proposed in [72] contains only 15 IGBTs as opposed to 18 IGBTs of the IMC and will be denoted as a sparse matrix converter (SMC) [102]. With a unidirectional PWM rectifier system the new converter can be realised with 9 IGBTs and 18 diodes and is the most simple form of the IMC. Due to the low number of power transistors the new topology is named as ultra sparse matrix converter (USMC) [72, 103]. In these topologies the indirect, sparse and ultra sparse matrix converter multistep switch commutation process is needed to reduce switching losses. The multistep commutation schemes used in the direct MC can be employed in those converters. In the next topology of IMC with reduced power switches, the diode bridge bidirectional switch cell arrangement is used at the source rectifier stage [114]. This switching configuration provides the zero DC link current commutation strategy [72], and such converters are called very sparse matrix converters (VSMC). Zero DC link current commutation also allows the employment of the circuit topology of the inverting link matrix converter (ILMC) [72]. Here, the bidirectional current carrying capability of the input stage is achieved by connecting, through two power transistors and two diodes, a conventional current DC link rectifier and a voltage inverter. Several kinds of unidirectional AC–AC topologies are presented in [115]. The topologies of the sparse, very sparse, ultra sparse and inverting link matrix converter also have a voltage transfer ratio of less than one, with a maximal level equal to 0.866.

With the rapid development of power electronics in high voltage, high power applications, the next steps in the development of direct power AC–AC converters introduces multilevel converters [52]. These converters are able to construct the output voltage waveforms with smaller voltage steps. Multilevel converter structures enable the voltage stress across the power semiconductor devices to be decreased with the increased number of voltage levels, enabling the use of medium voltage rated semiconductor devices to construct the converters for high voltage, high power applications. A multilevel indirect matrix converter (MIMC) is a topology that integrates the multilevel concept into the indirect matrix converter topology. In the literature are presented two main topologies, a three-level-output-stage indirect matrix converter [72, 83, 85] and an indirect three-level sparse matrix converter [82]. The first topology applies the three-level neutral-point-clamped voltage source inverter concept to the inversion stage of an indirect matrix converter topology. The second topology applies a simplified three-level neutral-point-clamped voltage source inverter concept with a neutral-point chopper to the inversion stage of an IMC topology. Having the ability to generate multilevel output voltages, both the MIMCs are able to produce better quality of output waveforms than a conventional IMC in terms of harmonic content. The disadvantages of such converters are the increase in the number of power switches and a more complicated modulation strategy.

A similar process takes place in the direct matrix converter. The multilevel direct matrix converter is obtained by replacing each switch in the classical direct MC by two or more series-connected switches and flying capacitors, which are introduced to clamp the voltage over the switches [104, 126]. In the three-level MC, two series-connected switches are introduced and six flying capacitors are connected to the mid-point of each of the two bi-directional switches. In a multilevel matrix converter, the voltage across the flying capacitor changes periodically following the input voltage. This kind of converter is not a classical direct AC–AC frequency converter without DC energy storage elements, because the flying capacitors are used as a local energy storage element. This direct multilevel MC may be applied to a high voltage range, with a lower level of harmonic contents on load voltage and source current compared with a conventional matrix converter [121].

The maximum voltage transfer ratio in both kinds of multilevel matrix converters is less than one. In a direct multilevel MC with Venturini modulation it is equal to 0.5 [104, 126], and with SVM it is equal to 0.8 [84, 101, 124]. In indirect multilevel MCs output voltages are also equal to 0.866 times source voltages [82, 83].

In order to obtain a voltage gain greater than one, several hybrid solutions have been proposed. The hybrid solution combines the direct AC–AC frequency converters and one small or several small local DC energy storage elements or DC–DC buck-boost (boost) choppers. The first topology was presented by Erickson and Al-Naseem in [31], and was named modular matrix converter (MMC) [6]. The MMC is obtained by replacing each switch in the classical direct MC [112] by a single phase H-bridge inverter. This topology has no main DC energy storage elements but there are several local small DC energy storage elements. This approach has the advantages of reduced switching loss and reduced harmonic content of output AC waveforms. The peak voltages applied to the semiconductor devices are clamped to capacitors and total switching loss is reduced. Furthermore, this converter can both increase and decrease the voltage amplitude and can operate with arbitrary power factors. The disadvantages are the complex power structures (high number of power semiconductor components and capacitors) and complex control strategy—each DC voltage in the capacitors has to be controlled via feedback.

The concept of a hybrid MC is presented in [66], where solutions with auxiliary H-bridge inverters integrated with MC or auxiliary H-bridge introduced to indirect MC are proposed. Both hybrid topologies provide improved transfer voltage, even equal to or higher than one. Another hybrid solution with integrated auxiliary voltage source, such as the DC–DC buck-boost or boost converter, at the intermediate DC link is presented in [69]. Several of this kind of topology are also proposed in paper [67]. Unity voltage transfer is also obtained in this solution [73, 74].

The next group of AC–AC frequency converters without DC energy storage with buck-boost voltage transformation, contains topologies based on PWM AC matrix-reactance choppers (MRC) structures [32, 38, 39] and direct matrix converter [112]. This group of frequency converters is the main object of publications. The first study of these kinds of converters was presented in 1993 by Antic et al. in [7–9]. The proposed topology was based on a three-phase buck-boost AC matrix-reactance chopper with source switches connected as in a matrix converter [112]. The authors presented the basic configuration of its power circuit and showed the first control theory for low speed induction motor drives. Later, in 2000, Zinoviev et al. continued the research on the presented MRFC topology. The detailed results are shown in a series of papers [89, 131, 132]. Then, the idea of connecting all of the unipolar PWM AC matrix-reactance choppers with a direct matrix converter was shown by Fedyczak in [32]. Later, Fedyczak called such converters matrix-reactance frequency converters (MRFC) [37]. The next series of papers was concerned with the MRFC based on the buck-boost MRC [35–37, 40–42, 45, 75, 107, 108] and the novel three structures of MRFC, based on Zeta [34, 36, 41], Ćuk [33, 36, 41], and boost [76] MRC topologies. Generally, in the presented MRFC, one with two groups of switches (source or load), it is connected as in direct MC. This approach makes it possible to obtain the load output voltage much greater than the source voltage. Its control was by modified Venturini modulation [112]. The conception of the MRFC with Venturini modulation was continuously developed by the authors and in the papers [43, 44, 107] the generation concept of a whole family of MRFCs based on unipolar PWM AC MRC is presented. The presented family of MRFC technologies employs a wide variety of topologies and contains nine topologies based on buck-boost, Ćuk, Zeta, SEPIC or boost MRC structures. It should be noted that MRFC can be controlled with all the control strategies used in a matrix converter, as can also be used the same commutation strategies.

Another concept of frequency converter topology with MRC was presented by Itoch in [58]. The presented topology is a cascade connection of the MRC with buck-boost topology and direct MC and is called a cascaded matrix converter. This topology can generally operate with reduced passive components as in MRFC, but there is a greater number of switches than for an MRFC.

It should be noted that in both MRFC and cascade connection, MRC and MC solutions electrical energy stored in the reactance elements during the period of input or output frequency is equal to zero, which is an essential difference in comparison to hybrid MC. Thus, the capacitors and reactors in an MRFC can be smaller. The voltage transfer ratio in both solutions is much greater than one. In recent time these topologies are developed in the literature as alternative solutions to hybrid converters [70, 71, 113].

In spite of several advantages of the frequency converters without DC energy storage, industrial application of such converters is still very limited because of some practical issues, such as low voltage transfer ratio. The development of matrix-reactance frequency converters seems to be one of several ways to increase the voltage transfer ratio in these frequency converters. The aim of this monograph is to make a concise presentation of selected frequency converters without DC energy storage and to give a detailed analysis of matrix-reactance frequency converters. In this book there will be presented selected research results related to topology generation, control strategies, commutation method, modelling, simulation and experimental verification of MRFC solutions. The book is recommended for Ph.D. students, scientists and engineers interested in issues related to power frequency converters without DC energy storage, especially fundamentals of matrix converters and matrix-reactance frequency converters, modelling of frequency converters and their practical implementation. Furthermore, the book constitutes also a review of frequency converter topologies without DC energy storage.

References

1. Agarwal V, Aggarwal RK, Patidar P, Patki C (2010) A Novel scheme for rapid tracking of maximum power point in wind energy generation systems. IEEE Trans Energy Convers 25(1):228–236
2. Ahmed SM, Iqbal A, Abu-Rub H, Rodriguez J, Rojas CA, Saleh M (2011) Simple carrier-based PWM technique for a three-to-nine-phase direct AC–AC converter. IEEE Trans Ind Electron 58(11):5014–5023
3. Ahmed SM, Iqbal A, Abu-Rub H (2011) Generalized duty-ratio-based pulsewidth modulation technique for a three-to-k phase matrix converter. IEEE Trans Ind Electron 58(9):3925–3937
4. Alesina A, Venturini M (1989) Analysis and design of optimum-amplitude nine-switch direct AC-AC converters. IEEE Trans Power Electron 4(1):101–112
5. Alesina A, Venturini M (1988) Intrinsic amplitude limits and optimum design of 9-switches direct PWM AC-AC converters. In: Proceedings of IEEE power electronics specialists conference, PESC'88, Kyoto, Japan, pp 1284–1291
6. Angkititrakul S, Erickson RW (2004) Control and implementation of a new modular matrix converter. In: Proceedings of IEEE applied power electronics conference and exposition, APEC'04, vol 2, Anaheim, US, pp 813–819
7. Antic D, Klaassens JB, Deleroi W (1993) A new power topology, suitable for low stator frequency operation of an induction machine. In: Proceedings of IEEE applied power electronics conference and exposition, APEC'93, San Diego, US, pp 146–152
8. Antic D, Klaassens JB, Deleroi W (1993) An integrated boost-buck and matrix converter topology for low speed drives. In: Proceedings of the EPE'93, Brighton, UK, pp 21–26

9. Antic D, Klaassens JB, Deleroi W (1994) Side effects in low-speed AC drives. In: Proceedings of IEEE power electronics specialists conference, PESC'94, Taipei, Taiwan, pp 998–1002

10. Arevalo SL, Zanchetta P, Wheeler PW, Trentin A, Empringham L (2010) Control and implementation of a matrix-converter-based AC ground power-supply unit for aircraft servicing. IEEE Trans Ind Electron 57(6):2076–2084

11. Barakati SM (2008) Applications of matrix converters for wind turbine systems. VDM Verlag, Berlin

12. Bekiarov A, Emadi SB (2002) Uninterruptible power supplies: classification, operation, dynamics, and control. In: Proceedings of IEEE applied power electronics conference and exposition, APEC'02, Dallas, US, pp 597–604

13. Benysek G (2007) Improvement in the quality of delivery of electrical energy using power electronics systems. Springer, London

14. Bhowmik S, Spée R (1993) A guide to the application-oriented selection of AC/AC converter topologies. IEEE Trans Power Electron 8(2):156–163

15. Blaabjerg F, Casadei D, Klumpner C, Matteini M (2002) Comparison of two current modulation strategies for matrix converters under unbalanced input voltage conditions. IEEE Trans Ind Electron 49(2):289–296

16. Boldea I, Nasar SA (2006) Electric drives, 2nd edn. CRC press, Boca Raton

17. Bose BK (2002) Modern power electronics and AC drives. Prentice Hall PTR, Upper Saddle River

18. Burany N (1989) Safe control of four-quadrant switches. In: Conference record of the IEEE industry applications conference annual meeting, IAS'89, pp 1190–1194

19. Cardenas R, Pena R, Clare J, Wheeler P (2011) Analytical and experimental evaluation of a WECS based on a bage induction generator fed by a matrix converter. IEEE Trans Energy Convers 26(1):204–215

20. Casadei D (2005) Tutorial on matrix converters. In: Proceedings of power electronics and intelligent control for energy conservation conference, PELINCEC'05, Warsaw, Poland

21. Casadei D, Grandi G, Serra G, Tanti A (1993) Space vector control of matrix converters with unity input power factor and sinusoidal input/output waveforms. In: Proceedings of European conference on power electronics and applications, EPE'93, vol 7, Brighton, UK, pp 170–175

22. Casadei D, Serra G, Tani A (1998) Reduction of the input current harmonic content in matrix converters under input/output unbalance. IEEE Trans Ind Electron 45(3):401–411

23. Casadei D, Serra G, Tani A, Nielsen P (1995) Performance of SVM controlled matrix converter with input and output unbalanced conditions. In: Proceedings of European conference on power electronics and applications, EPE'95, vol 2, Seville, Spain, pp 628–633

24. Casadei D, Serra G, Tani A, Zarri L (2009) Optimal use of zero vectors for minimizing the output current distortion in matrix converters. IEEE Trans Ind Electron 56(2):326–336

25. Casadei D, Serra G, Tanti A, Zaroi L (2002) Matrix converter modulation strategies: a new general approach based on space-vector representation of switch state. IEEE Trans Ind Electron 49(2):370–381

26. Casadei D, Trentin A, Matteini M, Calvini M (2003) Matrix converter commutation strategy using both output current and input voltage sign measurement. In: Proceedings of European conference on power electronics and applications, EPE'03, Toulouse, France, pp P1–P10 (CD-ROM)

27. Chakraborty S, Kramer B, Kroposki B (2009) A review of power electronics interfaces for distributed energy systems towards achieving low-cost modular design. Renew Sustain Energy Rev 13:2323–2335

28. Empringham L, de Lillo L, Khwan-On S, Brunson C, Wheeler PW, Clare JC (2011) Enabling technologies for matrix converters in aerospace applications. In: Proceedings of international conference—workshop compatibility and power electronics, CPE'2011, Tallinn, Estonia, pp 451–456

29. Empringham L, Wheeler PW, Clare JC (1998) Bi-directional switch current commutation for matrix converter applications. In: Proceedings of PE matrix converter, Prague, Czech Republic, pp 42–47

30. Empringham L, Wheeler PW, Clare JC (1998) Intelligent commutation of matrix converter bi-directional switch cells using novel gate drive techniques. In: Proceedings of power electronics specialists conference, PESC'98, Fukuoka, Japan, pp 707–713
31. Erickson RW, Al-Naseem OA (2001) A new family of matrix converters. In: Proceedings of IEEE industrial electronics society conference, IECON'01, vol 2, Denver, US, pp 1515–1520
32. Fedyczak Z (2003) PWM AC voltage transforming circuits (in Polish). Zielona Góra University Press, Zielona Góra
33. Fedyczak Z, Szcześniak P (2006) Koncepcja matrycowo-reaktancyjnego przemiennika częstotliwości typu Ćuk (in Polish). Przegląd Elektrotechniczny (Electr Rev) 7/8:42–47
34. Fedyczak Z, Szcześniak P (2006) Koncepcja matrycowo-reaktancyjnego przemiennika częstotliwości typu Zeta (in Polish). Wiadomości Elektrotechniczne (Electrotech News) 3:26–29
35. Fedyczak F, Szcześniak P (2009) Modelling and analysis of matrix-reactance frequency converters using voltage source matrix converter and LF transfer matrix modulation method. Przegląd Elektrotechniczny (Electr Rev) 2:125–130
36. Fedyczak Z, Szcześniak P (2007) New matrix-reactance frequency converters—conception description. In: Orłowska-Kowalska T (ed) Power electronics and electrical drives: selected problems. Wrocław Technical University Press, Wrocław, pp 71–84
37. Fedyczak Z, Szcześniak P (2005) Study of matrix-reactance frequency converter with buck-boost topology. In: Proceedings of power electronics and intelligent control for energy conservation conference, PELINCEC'05, Warsaw, Poland (CD-ROM)
38. Fedyczak Z, Klytta M, Strzelecki R (2001) Three-phase AC/AC semiconductor transformer topologies and applications. In: Proceedings of power electronics devices compatibility conference, PEDC'01, Zielona Góra, Poland, pp 25–38
39. Fedyczak Z, Strzelecki R, Sozański K (2002) Review of three-phase AC/AC semiconductor transformer topologies and applications. In: Proceedings of symposium on power electronics, electrical drives automation and motion, SPEEDAM'02, Ravello, Italy, pp B.5-19–B.5-24
40. Fedyczak Z, Szcześniak P, Jankowski M (2005) Koncepcja matrycowo-reaktancyjnego przemiennika częstotliwości typu buck-bost (in Polish). Sterowanie w Energoelektronice i Napędzie Elektrycznym, SENE'05, number 1, Łódź, Poland, pp 101–106
41. Fedyczak Z, Szcześniak P, Kaniweski J (2007) Direct PWM AC choppers and frequency converters. In: Korbicz J (ed) Measurements models systems and design. Transport and Communication Publishers, Warsaw, pp 393–424
42. Fedyczak Z, Szcześniak P, Klytta M (2006) Matrix-reactance frequency converter based on buck-boost topology. In: Proceedings of power electronics and motion control conference, EPE-PEMC'06, Portoroz, Slovenia, pp 763–768
43. Fedyczak Z, Szcześniak P, Korotyeyev I (2008) Generation of matrix-reactance frequency converters based on unipolar PWM AC matrix-reactance choppers. In: Proceedings of IEEE power electronics specialists conference, PESC'08, Rhodes, Greece, pp 1821–1827
44. Fedyczak Z, Szcześniak P, Korotyeyev I (2008) New family of matrix-reactance frequency converters based on unipolar PWM AC matrix-reactance choppers. In: Proceedings of power electronics and motion control conference, EPE-PEMC'08, Poznań, Poland, pp 236–243
45. Fedyczak Z, Szcześniak P, Kaniweski J, Tadra G (2009) Implementation of three-phase frequency converters based on PWM AC matrix-reactance chopper with buck-boost topology. In: Proceedings of European conference on power electronics and applications, EPE'09, Barcelona, Spain, pp P1–P10 (CD-ROM)
46. Fedyczak Z, Tadra G, Klytta M (2010) Implementation of the current source matrix converter with space vector modulation. In: Proceedings of power electronics and motion control conference, EPE-PEMC'10, Ohrid, Macedonia (CD-ROM)
47. Gyugi L, Pelly B (1976) Static power frequency changers: theory, performance and applications. Wiley, New York
48. Helle L, Munk-Nielsen S (2001) A novel loss reduced modulation strategy for matrix converters. in: Proceedings of IEEE power electronics specialists conference, PESC'01, vol 2, Vancouver, Canada, pp 1102–1107

49. Helle L, Larsen KB, Jorgensen HA, Munk-Nielsen S (2004) Evaluation of modulation schemes for three-phase to three-phase matrix converters. IEEE Trans Ind Electron 51(1):158–171
50. Hingorani N, Gyugyi L (2000) Understanding FACTS: concepts and technology of flexible AC transmission systems. IEEE, New York
51. Hofmann W, Ziegler M (2001) Multi-step commutation and control policies for matrix converters. In: Proceedings of international conference on power electronics, ISPE'01, Seoul, Korea, pp 795–802
52. Holmes DG, Lipo TA (2003) Pulse width modulation for power converters. Principle and practice. IEEE Press, New York
53. Huber L, Borojević D (1995) Space vector modulated three-phase to three-phase matrix converter with input power factor correction. IEEE Trans Ind Appl 31(6):1234–1246
54. Huber L, Borojević D, Burany N (1989) Voltage space vector based PWM control of forced commutated cycloconverters. In: Proceedings of industrial electronics society annual conference, IECON'89, vol 1, pp 106–111
55. Iimori K, Shinohara K, Yamamoto K (2006) Study of dead time of PWM rectifier of voltage-source inverter without DC-link components and its operating characteristics of induction motor. IEEE Trans Ind Appl 42(2):518–525
56. Iimori K, Shinohara K, Tarumi O, Fu Z, Muroya M (1997) New current-controlled PWM rectifier voltage source inverter without DC-link components. In: Proceedings of power conversion conference, PCC'97, vol 2, Nagaoka, Japan, pp 783–786
57. Ishiguro A, Furuhashi T, Okuma S (1991) A novel control method for forced commutated cycloconverters using instantaneous values of input line-to-line voltages. IEEE Trans Ind Electron 38(3):166–172
58. Itoh J-I, Koiwa K, Kato K (2010) Input current stabilization control of a matrix converter with boost-up functionality. In: Proceedings of international power electronics conference, IPEC 2010, Sapporo, Japan
59. Jussila M, Tuusa H (2007) Comparison of simple control strategies of space-vector modulated indirect matrix converter under distorted supply voltage. IEEE Trans Power Electron 22(1):139–148
60. Jussila M, Salo M, Tuusa H (2003) Realization of a three-phase indirect matrix converter with an indirect vector modulation method. In: Proceedings of power electronics specialist conference, PESC'03, vol 2, Acapulco, Meksyk, pp 689–694
61. Kaźmierkowski MP, Krishnan R, Blaabjerg F (2002) Control in power electronics: selected problems. Academic Press Series in Engineering, New York
62. Keyhani A, Marwali MN, Dai M (2009) Integration of green and renewable energy in electric power systems. Wiley, New York
63. King AC, Knight W (2003) Uninterruptible power supplies and standby power systems. McGraw-Hill, New York
64. Klumpner C (2005) Hybrid direct power converters with increased/higher than unity voltage transfer ratio and improved robustness against voltage supply disturbances. In: Proceedings of power electronics specialists conference, PESC'05, pp 2383–2389
65. Klumpner C, Blaabjerg F (2003) Two-stage direct power converters: an alternative to matrix converters. In: IEE matrix converter seminar, Birmingham, UK
66. Klumpner C, Pitic C (2008) Hybrid matrix converter topologies: an exploration of benefits. In: Proceedings of power electronics specialists conference, PESC'08, Rhodes, Greece, pp 2–8
67. Klumpner C, Blaabjerg F, Thogersen P (2003) Converter topologies with low passive components usage for the next generation of integrated motor drives. In: Proceedings of power electronics specialist conference, PESC'03, vol 2, pp 568–573
68. Klumpner C, Nielsen P, Boldea I, Blaabjerg F (2002) A new matrix converter motor (MCM) for industry applications. IEEE Trans Ind Electron 49(2):325–335
69. Klumpner C, Wijekoon T, Wheeler P (2005) A new class of hybrid AC/AC direct power converters. In: Proceedings of IAS annual meeting industry applications conference, IAS'05, vol 4, Hong Kong, pp 2374–2381

70. Koiwa K, Itoh J-I (2011) A gain design method of a damping control for a matrix converter. In: 2011 annual meeting IEEJ, Toyonaka-city, Osaka, Japan, pp 1–2
71. Koiwa K, Itoh J-I (2011) Experimental verification for a matrix converter with a V-connection AC chopper. In: Proceedings of European conference on power electronics and applications, EPE'11, Birmingham, UK, pp 1–10
72. Kolar JW, Baumann M, Schafmeister F, Ertl H (2002) Novel three-phase AC-DC-AC sparse matrix converter. In: Proceedings of IEEE applied power electronics conference and exposition, APEC'02, vol 2, Dallas, US, pp 777–791
73. Kolar JW, Friedli T, Krismer F, Round SD (2008) The essence of three-phase AC/AC converter systems. In: Proceedings of power electronics and motion control conference, EPE-PEMC'08, Poznań, Poland, pp 27–42
74. Kolar JW, Friedli T, Rodriguez J, Wheeler PW (2011) Review of three-phase PWM AC-AC converter topologies. IEEE Trans Ind Electron 58(11):4988–5006
75. Korotyeyev I, Fedyczak Z (2008) Steady and transient states modelling methods of matrix-reactance frequency converter with buck-boost topology. COMPEL (Int J Comput Math Electr Electron Eng) 28(3):626–638
76. Korotyeyev I, Fedyczak Z, Szcześniak P (2008) Steady and transient state analysis of a matrix-reactance frequency converter based on a boost PWM AC matrix-reactance chopper. In: Proceedings of the international school on nonsinusoidal currents and compensation, ISNCC'08, Łagów, Poland (CD-ROM)
77. Kwon BH, Min B-D, Kim J-H (1998) Novel commutation technique of AC-AC converters. IEE Proc Electr Power Appl 145(4):295–300
78. Kwon WH, Cho GH (1993) Analyses of static and dynamic characteristics of practical step-up nine-switch convertor. IEE Proc-B 140(2):139–145
79. Kwon WH, Cho GH (1991) Analysis of non-ideal step down matrix converter based on circuit DQ transformation. In: Proceedings of power electronics specialists conference, PESC'91, Cambridge, US, pp 825–829
80. Lai R, Wang F, Burgos R, Pei Y, Boroyevich D, Lipo TA, Immanuel VD, Karimi KJ (2008) A systematic topology evaluation methodology for high-density three-phase PWM AC-AC converters. IEEE Trans Power Electron 23(6):2665–2680
81. Lee KB, Blaabjerg F (2008) Simple power control for sensorless induction motor drives fed by a matrix converter. IEEE Trans Energy Convers 23(3):781–788
82. Lee MY, Klumpner C, Wheeler PW (2008) Experimental evaluation of the indirect three-level sparse matrix converter. In: Proceedings of IET international conference on power electronics, machines and drives, PEMD'08, York, UK, pp 50–54
83. Lee MY, Wheeler PW, Klumpner C (2007) Modulation method for the three-level-output-stage matrix converter under balanced and unbalanced supply condition. In: Proceedings of European conference on power electronics and applications, EPE'07, Alborg, Denmark, pp 1–10 (CD-ROM)
84. Lie X, Clare JC, Wheeler PW, Empringham L (2008) Space vector modulation for a capacitor clamped multi-level matrix converter. In: Proceedings of power electronics and motion control conference, EPE-PEMC'08, Poznań, Poland, pp 229–235
85. Loh PC, Blaabjerg F, Gao F, Baby A, Tan DA (2008) Pulsewidth modulation of neutral-point-clamped indirect matrix converter. IEEE Trans Ind Appl 44(6):1805–1814
86. Mahlein J, Igney J, Braun M, Simon O (2001) Robust matrix converter commutation without explicit sign measurement. In: Proceedings of European conference on power electronics and applications, EPE'01 (CD-ROM)
87. Nguyen HM, Lee H-H, Chun T-W (2011) Input power factor compensation algorithms using a new direct-SVM method for matrix converter. IEEE Trans Ind Electron 58(1):232–243
88. Nielsen P, Blaabjerg F, Pedersen JK (1996) Space vector modulated matrix converter with minimized number of switchings and a feedforward compensation of input voltage unbalance. in: Proceedings of international power electronics, drives and energy systems for industrial, growth, PEDES'96, vol 2, pp 833–839

89. Obuchov AY, Otchenasch W, Zinoviev GS (2000) Buck-boost AC-AC voltage controllers. In: Proceedings of international conference on power electronics and motion control, EPE-PEMC 2000, Košice, Slovakia, pp 2.194–2.197
90. Ortega C, Arias A, Caruana C, Balcells J, Asher GM (2010) Improved waveform quality in the direct torque control of matrix-converter-fed PMSM drives. IEEE Trans Ind Electron 57(6):2101–2110
91. Oyama J, Xia X, Higuchi T, Yamada E (1997) Displacement angle control of matrix converter. In: Proceedings of IEEE power electronics specialists conference, PESC'97, St. Louise, US, pp 1033–1039
92. Patel H, Hoft RG (1973) Generalized techniques of harmonic elimination and voltage control in thyristor inverters: Part I, Harmonic elimination. IEEE Trans Ind Electron 3:310–317
93. Patel H, Hoft RG (1974) Generalized techniques of harmonic elimination and voltage control in thyristor inverters: Part II, Voltage control techniques. IEEE Trans Ind Electron IA-1O(5):666–673
94. Pinto FS, Silva FJ (1999) Sliding mode control of space vector modulated matrix converter with sinusoidal input/output waveforms and near unity input power factor. In: Proceedings of European conference on power electronics and applications, EPE'99, Lausanne, Switzerland, pp 1–9
95. Podlesak TF, Katsis D, Wheeler PW, Clare J, Empringham L, Bland M (2005) A 150-kVA vector controlled matrix converter induction motor drive. IEEE Trans Ind Appl 41(3):841–847
96. Rodriguez J (1983) A new control technique for AC-AC converters. In: Proceedings of control in power electronics and electrical drives conference, IFAC'83, Lausanne, Switzerland, pp 203–208
97. Rodriguez J, Rivera M, Kolar JW, Wheeler PW (2012) A review of control and modulation methods for matrix converters. IEEE Trans Ind Electron 59(1):58–70
98. Roy G, April GE (1989) Cycloconverter operation under a new scalar control algorithm. In: Proceedings of power electronics specialists conference, PESC'89, vol 1, Milwaukee, US, pp 368–375
99. Roy G, April GE (1991) Direct frequency changer operation under a new scalar control algorithm. IEEE Trans Power Electron 6(1):100–107
100. Roy G, Duguay L, Manias S, April GE (1987) Asynchronous operation of cycloconverter with improved voltage gain by employing a scalar control algorithm. In: Proceedings of IEEE-IAS annual meeting, pp 889–898
101. Rząsa J (2007) Wielopoziomowy przekształtnik matrycowy sterowany metodą venturiniego (in Polish). Przegląd Elektrotechniczny (Electr Rev) 2:57–64
102. Schafmeister F, Baumann M, Kolar JW (2002) Analytically closed calculation of the conduction and switching losses of three-phase AC-AC sparse matrix converters. In: Proceedings of international power electronics and motion control conference, EPE-PEMC'02, Dubrovnik, Croatia, pp 1–13 (CD-ROM)
103. Schonberger J, Friedli T, Round SD, Kolar JW (2007) An ultra sparse matrix converter with a novel active clamp circuit. In: Proceedings of power conversion conference, PCC'07, Nagoya, Japan, pp 784–791
104. Shi Y, Yang X, He Q, Wang Z (2004) Research on a novel multilevel matrix converter. In: Proceedings of IEEE power electronics specialists conference, PESC'04, vol 3, Aachen, Germany, pp 2413–2419
105. Strzelecki R, Benysek G (2008) Power electronics in smart electrical energy networks. Springer, London
106. Sun K, Zhou D, Huang L, Matsuse K, Sasagawa K (2007) A novel commutation method of matrix converter fed induction motor drive using RB-IGBT. IEEE Trans Ind Appl 43(3):777–786
107. Szcześniak P (2009) Analysis and testing matrix-reactance frequency converters. PhD thesis (in Polish), University of Zielona Góra, Zielona Góra

108. Szcześniak P, Fedyczak Z, Klytta M (2008) Modelling and analysis of a matrix-reactance frequency converter based on buck-boost topology by DQ0 transformation. In: Proceedings of power electronics and motion control conference, EPE-PEMC'08, Poznań, Poland, pp 165–172

109. Teodorescu R, Liserre M, Rodriguez P (2011) Grid converters for photovoltaic and wind power systems. Wiley-IEEE, New York

110. Vargas R, Ammann U, Rodriguez J, Pontt J (2008) Predictive strategy to control common-mode voltage in loads fed by matrix converters. IEEE Trans Ind Electron 55(12):4372–4380

111. Vargas R, Ammann U, Hudoffsky B, Rodriguez J, Wheeler P (2010) Predictive torque control of an induction machine fed by a matrix converter with reactive input power control. IEEE Trans Power Electron 25(6):1426–1438

112. Venturini M, Alesina A (1980) The generalized transformer: a new bi-directional sinusoidal waveform frequency converter with continuously adjustable input power factor. In: Proceedings of IEEE power electronics specialists conference, PESC'80, pp 242–252

113. Wan HC, Wen BC, Qing AC, Wang Y (2010) The research of new topology of matrix converter with high voltage transfer ratio based on "pump-type" structure. Appl Electron Tech 36(5):87–90

114. Wei L, Lipo TA (2001) A novel matrix converter topology with simple commutation. In: Proceedings of IEEE industry applications society annual meeting, IAS'01, vol 3, Chicago, US, pp 1749–1754

115. Wei L, Lipo TA, Chan H (2002) Matrix converter topologies with reduced number of switches. In: Proceedings of power electronics specialists conference, PESC'02, vol 1, Cairns, Australia, pp 57–63

116. Wheeler PW, Clare JC, Apap M, Bradley KJ (2008) Harmonic loss due to operation of induction machines from matrix converters. IEEE Trans Ind Electron 55(2):809–816

117. Wheeler PW, Clare JC, Apap M, Empringham L, Bradley KJ, Pickering S, Lampard DA (2005) Fully integrated 30 kW motor drive using matrix converter technology. In: Proceedings of European conference on power electronics and applications, EPE'05, Dresden, pp 2390–2395

118. Wheeler PW, Clare J, Empringham L (2004) Enhancement of matrix converter output waveform quality using minimized commutation times. IEEE Trans Ind Electron 51(1):240–244

119. Wheeler PW, Empringham L, Clare J (2002) Minimization of matrix converter commutation times. In: Proceedings of power electronics and motion control conference—EPE-PEMC'02, Dubrovnik, Croatia (CD-ROM)

120. Wheeler PW, Rodriguez J, Clare JC, Empringham L, Weinstejn A (2002) Matrix converters: a technology review. IEEE Trans Ind Electron 49(2):276–288

121. Wheeler PW, Lie X, Lee MY, Empringham L, Klumpner C, Clare J (2008) A review of multi-level matrix converter topologies. In: Proceedings of IET international conference on power electronics, machines and drives, PEMD'08, York, UK, pp 286–290

122. Wijekoon T, Klumper C, Zanchetta P, Wheeler PW (2008) Implementation of a hybrid AC-AC direct power converter with unity voltage transfer. IEEE Trans Power Electron 23(4):1918–1926

123. Xiao D, Rahman FM (2010) Implementation of sensorless direct torque control using matrix converter fed Interior permanent magnet synchronous motor. In: International power electronics conference, IPEC'2010, Sapporo, Japan, pp 3065–3071

124. Xu L, Clare JC, Wheeler PW, Empringham L, Li Y (2012) Capacitor clamped multilevel matrix converter space vector modulation. IEEE Trans Ind Electron 59(1):105–115

125. Yamamoto E, Hara H, Kang JK, Krug HP (2011) Development of MCs for industrial applications. IEEE Indus Electron Mag 5:4–12

126. Yang X, Shi Y, He Q, Wang Z (2004) A novel multi-level matrix converter. In: Proceedings of IEEE applied power electronics conference and exposition, APEC'04, vol 2, Anaheim, US, pp 832–835

127. Yoon Y-D, Sul S-K (2006) Carrier-based modulation technique for matrix converter. IEEE Trans Power Electron 21(6):1691–1703

128. Ziegler M, Hofmann W (2000) A new two steps commutation policy for low cost matrix converters. In: Proceedings of PCIM conference, Nürnberg, Germany
129. Ziegler M, Hofmann W (2001) New one-step commutation strategies in matrix converters. In: Proceedings of power electronics and drive systems conference, PEDS'01, vol 2, Bali, Indonesia, pp 560–564
130. Ziegler M, Hofmann W (1998) Semi natural two steps commutation strategy for matrix converters. In: Proceedings of power electronics specialists conference, PESC'98, Fukuoka, Japan, pp 727–731
131. Zinoviev GS, Obuchov AY, Otchenasch WA, Popov WI (2000) Transformerless PWM AC boost and buck-boost converters (in Russian). Technicznaja Elektrodinamika 2:36–39
132. Zinoviev GS, Ganin M, Levin E, Obuchov AY, Popov V (2000) New class of buck-boost AC-AC frequency converters and voltage controllers. In: Proceedings of Korea-Russia international symposium on science and technology, KORUS'2000, Ulsan, Korea, pp 303–308
133. Ziogas PD, Khan SI, Rashid MH (1986) Analysis and design of forced commutated cyclo-converter structures with improved transfer characteristics. IEEE Trans Ind Electron IE-33: 271–280
134. Ziogas PD, Khan SI, Rashid MH (1985) Some improved forced commutated cycloconverters structures. IEEE Trans Ind Appl 1A-21:1242–1253

Chapter 2
Review of AC–AC Frequency Converters

2.1 Introduction

As mentioned above, frequency converters convert AC electrical power of one frequency into AC electrical power of another frequency [51]. Additionally, this kind of converter also has the capability to control the load voltage amplitude, the load displacement angle relative to source voltage, the displacement angle between source currents and voltages (input power factor) and the capability to control bi-directional (or only unidirectional) power flow through the converter [51]. Figure 2.1 shows a generic three-phase PWM AC–AC frequency converter diagram and functional representation of such frequency converters.

To the input terminal of the frequency converter are connected voltage sinusoidal AC sources, with constant amplitude U_S and constant frequency f_S. These applied voltages are converted into output voltage waves with set amplitude U_L, frequency f_L and displacement angle of the load voltages relative to source voltages L_S. These output voltages are applied to the load. The load current amplitudes I_L and phase angles φ_L are determined by the impedance characteristic of the loads. During bi-directional power flow control in the case of direction from output terminals to input terminals, the frequency converter converts the load current waves of frequency f_L, into input current waves of frequency f_S.

AC–AC frequency converter topologies can be broadly classified into three categories, depending upon the type of AC–AC conversion. Figure 2.2 shows a classification tree for frequency converters. The classification of AC–AC frequency converters in the technical literature is varied, because the development of the converters discussed is still in progress [16, 81, 83, 84, 126]. The latter classified as indirect structures with main DC energy storage elements, direct structures without DC energy storage element and hybrid structures with small local DC energy storage elements. The first group includes the most popular and widely used in industry and households, i.e. direct frequency converters with voltage source inverters (VSI) or

P. Szcześniak, *Three-Phase AC–AC Power Converters Based on Matrix Converter Topology*, Power Systems, DOI: 10.1007/978-1-4471-4896-8_2, © Springer-Verlag London 2013

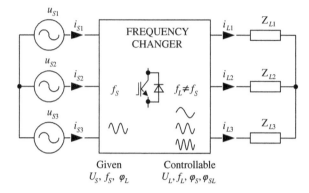

Fig. 2.1 Generic diagram of three-phase PWM AC–AC frequency converter and its functionality

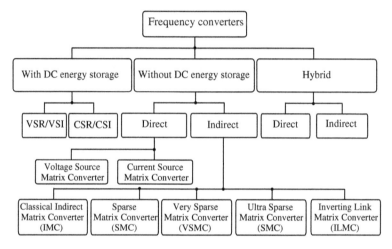

Fig. 2.2 Classification of three-phase AC–AC converter topologies

current source inverters (CSI). The second group consists of alternative topologies of direct frequency converters. These topologies have no DC energy storage elements and basically consist of an array of static power switches connected between the source and load terminals. For good performance direct frequency converters have small capacitors and inductors, such as high frequency component filters or small regenerative AC energy storage. The last group is a combination of direct frequency converters with small-sized local DC energy storage elements or an additional module with a DC–DC boost converter.

2.2 Frequency Converters with a DC Energy Storage Element

The most traditional AC–AC power converter topology is a pulse width modulated (PWM) voltage source inverter (PWM-VSI) with a front-end diode rectifier and a DC link capacitor, as shown in Fig. 2.3 [59, 72]. The frequency converter presented in Fig. 2.3 is also called a two-level indirect converter with voltage source inverter (VSI). An indirect converter consists of two converter stages and energy storage element, which convert input AC into DC and then reconvert DC back into output AC with variable amplitude and frequency. The DC-link capacitor decouples two AC power conversion stages and ensures the independent control of two stages. The control of the output is achieved by modulating the duty cycles of the devices in the inverter stage so as to produce near-sinusoidal output currents in the inductive load, at a desired amplitude and frequency. The source current in this converter is highly distorted, containing high amounts of low-order harmonics (5th and 7th) [11]. Through the impedance of the mains, the low-order current harmonics may distort the voltage at the point of common coupling, which may further interfere with other electric systems in the network. As the current direction in a diode rectifier cannot reverse, some mechanism must be implemented to handle an eventual energy flow reversal, such as during an electromagnetic braking of a motor, in order to prevent the DC bus voltage reaching destructive levels. Such mechanisms are always dissipative ones (in a braking resistor) and hence they can be effectively employed only when the energy to be dissipated is low [72]. The solution to the problem is to use an IGBT bridge as a supply rectifier. This converter is called back-to-back inverter (B2B VSI) and is presented in Fig. 2.4 [145]. The back-to-back converter consists simply of a force-commutated rectifier and a force-commutated inverter connected by a common DC-link, making their separate control possible. The line-side converter may be operated to give sinusoidal source currents, and the braking energy can be fed back to the power grid. It is a boost-type converter, i.e. its DC-link side voltage has to be higher than the peak value of the supply line-to-line voltage. In the back-to-back VSI, source filter inductors are also required. These inductors are a big problem

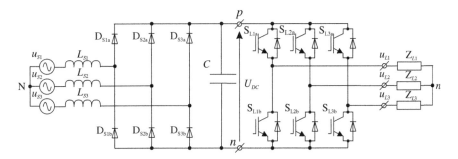

Fig. 2.3 Two-level indirect frequency converter with voltage source inverter and diode bridge rectifier

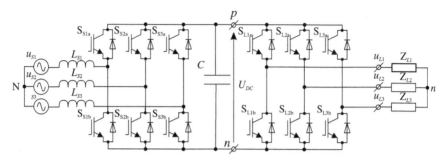

Fig. 2.4 Back-to-back converter (BB-VSI)

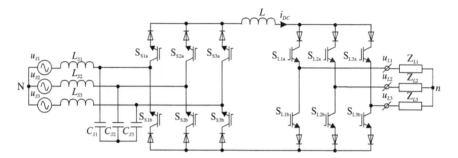

Fig. 2.5 Two-level indirect frequency converter with current source inverter

because the inductors are bulkier and heavier than the DC link capacitor in low and medium power converters.

An alternative solution to a frequency converter with voltage source inverter is the solution with PWM current source inverter (CSI), presented in Fig. 2.5 [34, 59]. The CSI produces sinusoidal supply current waveforms similar to the back-to-back VSI. The CSC contains a DC link inductor, which is generally larger and heavier than the link capacitor in voltage source converters. In the CSI, a source filter is also required. This is a low-pass LC-type filter, and the physical size of CSI source filter is smaller than used in a B2B VSI. Furthermore, the CSI usually requires series-connected diodes with every IGBT. This increases semiconductor conduction losses and the complexity of the main circuit.

The DC energy storage in the presented indirect frequency converters is a bulky component. In the solution with VSI the DC link capacitors are relatively large compared to the size of the rectifier and inverter semiconductor components, at the same time reducing the speed of response. Electrolytic capacitors typically occupy from 30 to 50 % of the total volume of the converter for power levels greater than a few kW and in addition to this they are a component with a limited lifetime. It should be noted that the electrolytic capacitor has by far the shortest lifetime of any element, active or passive, used in power electronic converters. In addition, the presence of the capacitor significantly limits the power converter to high temperature applications up to 300 °C, because these capacitors are temperature sensitive. Capacitors also cause higher maintenance costs of the conversion system. Furthermore, high power conventional capacitors cannot be used in some special applications, such as in aeronautics, aircraft and deep-sea or space systems [12, 15, 19, 31, 141]. In the case of the CSI, the DC-link inductors are generally bulkier and heavier than the link capacitor in voltage source converters. However, frequency converters with VSI, (diode rectifier stage and back-to-back) are well known and widely used in industry.

The classical VSI generates a low-frequency output voltage with controllable magnitude and frequency by programming high-frequency voltage pulses. Of the various pulse-programming methods, the carrier-based pulse width modulation (PWM) methods are the preferred approach in most applications [52]. Two main implementation techniques exist in the control of load voltages: The first is the triangle intersection technique, where the reference modulation wave is compared with a triangular carrier wave and the intersections define the switching instants. The second is based on space vector modulation (SVM). In this method, the time length of the inverter states are precalculated for each carrier cycle by employing space-vector theory [50, 132].

In the classical three-phase VSI stage in the frequency converter shown in Figs. 2.3 and 2.4 is identified as eight switch combinations, which are connected in Table 2.1. Two of these states are a short circuit of the output terminals while the other six produce active voltages. In the three-phase CSI stage shown in Fig. 2.5, there is possible six active switch combinations and three with zero output current. The switch combinations in CSI are presented in Table 2.2.

In the space-vector approach, employing the complex variable transformation, the time domain load voltage and current signals are translated to the complex reference voltage or current vector, which rotates in the complex coordinates with the angular speed shown in the following:

$$\underline{\mathbf{u}}_L = \frac{2}{3}(u_{L1} + \underline{a}u_{L2} + \underline{a}^2 u_{L3}), \tag{2.1}$$

$$\underline{\mathbf{i}}_L = \frac{2}{3}(i_{L1} + \underline{a}i_{L2} + \underline{a}^2 i_{L3}), \tag{2.2}$$

Table 2.1 Switch configurations and corresponding load voltages in VSI

No.	S_{L1a}	S_{L2a}	S_{L3a}	S_{L1b}	S_{L2b}	S_{L3b}	u_{L1n}	u_{L2n}	u_{L3n}
1	1	1	0	0	0	1	$\frac{1}{3}U_{DC}$	$\frac{1}{3}U_{DC}$	$-\frac{2}{3}U_{DC}$
2	1	0	0	0	1	1	$\frac{2}{3}U_{DC}$	$-\frac{1}{3}U_{DC}$	$-\frac{1}{3}U_{DC}$
3	0	1	0	1	0	1	$-\frac{1}{3}U_{DC}$	$\frac{2}{3}U_{DC}$	$-\frac{1}{3}U_{DC}$
4	0	1	1	1	0	0	$-\frac{2}{3}U_{DC}$	$\frac{1}{3}U_{DC}$	$\frac{1}{3}U_{DC}$
5	0	0	1	1	1	0	$-\frac{1}{3}U_{DC}$	$-\frac{1}{3}U_{DC}$	$\frac{2}{3}U_{DC}$
6	1	0	1	0	1	0	$\frac{1}{3}U_{DC}$	$-\frac{2}{3}U_{DC}$	$\frac{1}{3}U_{DC}$
7	1	1	1	0	0	0	0	0	0
8	0	0	0	1	1	1	0	0	0

Table 2.2 Switch configurations and corresponding load current in CSI

No.	S_{L1a}	S_{L2a}	S_{L3a}	S_{L1b}	S_{L2b}	S_{L3b}	i_{L1}	i_{L2}	i_{L3}
1	1	0	0	0	0	1	I_{DC}	0	$-I_{DC}$
2	0	1	0	0	0	1	0	I_{DC}	$-I_{DC}$
3	0	1	0	1	0	0	$-I_{DC}$	I_{DC}	0
4	0	0	1	1	0	0	$-I_{DC}$	0	I_{DC}
5	0	0	1	0	1	0	0	$-I_{DC}$	I_{DC}
6	1	0	0	0	1	0	I_{DC}	$-I_{DC}$	0
7	1	0	0	1	0	0	0	0	0
8	0	1	0	0	1	0	0	0	0
9	0	0	1	0	0	1	0	0	0

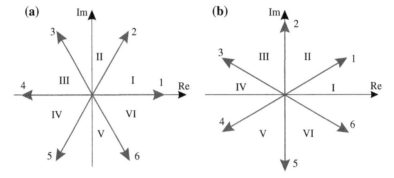

Fig. 2.6 Active stationary vector on dq plane for three-phase: **a** VSI (Table 2.1), **b** CSI (Table 2.2)

where: $\underline{a} = e^{j\frac{2\pi}{3}}$, $\mathbf{u}_L$, $\mathbf{i}_L$-vectors of load voltages and load current, respectively; U_L, I_L-magnitude of those vectors; α_L, β_L-phase angle of load voltage and load current vectors, respectively. Each active switch configuration in VSI and CSI corresponds to active space vectors, while the zero configuration corresponds to zero space vectors. Active vectors are represented in the dq plane as shown in Fig. 2.6, and spaced equally at 60° intervals around the complex plane [59].

2.3 Frequency Converters Without DC Energy Storage Element

2.3.1 Introduction

The main aim of this subsection is to describe the general characteristics of AC–AC frequency converters without DC energy storage elements. As is presented in Fig. 2.2 these converters are divided into three groups. The first group contains a classical direct matrix converter which operates in two modes: as a voltage source matrix converter and a current source matrix converter, similarly as in converters with DC energy storage elements. The second group contains indirect converters with fictitious DC-link (but without DC storage elements). The third group is converters based on matrix-reactance choppers with source or load synchronous switches connected as in a matrix converter. This concept is based on regenerative AC energy storage elements such as small capacitors or inductors. In these elements the average energy during the source (or load) voltage time period is equal to zero.

The main focus is given to the presented fundamentals of the matrix converter, which are the fundamental structures in AC–AC frequency converters without DC energy storage. Furthermore, matrix-frequency converters, as the main object of this book will be described in detail in the following chapters.

2.3.2 Direct AC–AC Frequency Converters: Matrix Converter

As mentioned above, the groups of direct frequency converters include matrix converter (MC) structures. The MC, depending on the kind of power supply (voltage or current character), can work as a voltage source matrix converter (VSMC) or a current source matrix converter (CSMC), respectively. The main technical papers presented at conferences and in journals concern the matrix converter in VSMC mode, and commonly this structure is referred to as a matrix converter.

Generally, the matrix converter is a single-stage converter which has an array of m × n bi-directional power switches to connect, directly, an m-phase voltage source to an n-phase load [1, 2, 135, 142], which is presented in Fig. 2.7. In three-phase systems, an MC is an array of nine bi-directional switches that allow any load phase to be connected to any source phase (Fig. 2.8). In the case of voltage sources on the input side there are voltage source matrix converters (VSMC), the schemes of which are depicted in Figs. 2.7 and 2.8 [87].

For good performance, the VSMC should have a source filter. The source filter is generally needed to minimise the high frequency components in the input currents and reduce the impact of the perturbations from the input grid. Their size is inversely proportional to the matrix converter switching frequency. The major advantage of matrix converters is the absence of the DC link capacitor, which may increase the efficiency and the lifetime of the converter. The development of MCs started when Venturini and Alesina proposed the basic principles of operation in the early 1980s [135]. The authors proposed in [135] a high switching frequency control algorithm and the development of a rigorous mathematical analysis to describe the low-frequency behaviour of the converter.

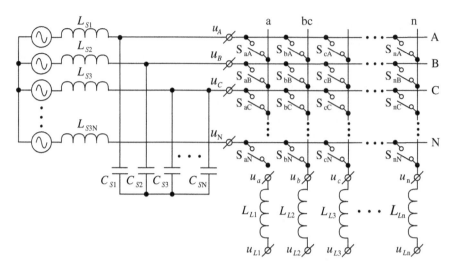

Fig. 2.7 Simplified circuit of m × n phase matrix converter

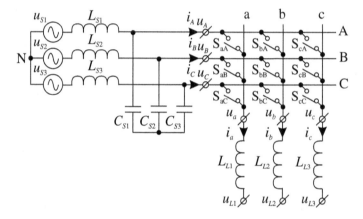

Fig. 2.8 Simplified circuit of three-phase matrix converter

Basic principles of matrix converter

The switching function of a single switch is defined as follows [135]:

$$s_{jK} = \begin{cases} 1, & \text{switch } s_{jK} \text{ turn-on} \\ 0, & \text{switch } s_{jK} \text{ turn-off} \end{cases}, \tag{2.3}$$

where $j \in \{a, b, c\}$ is the name of the output phase, $K \in \{A, B, C\}$ is the name of the input phase. Taking into account that the input phases must never be short-circuited and that the output currents must never be interrupted, the constraints can

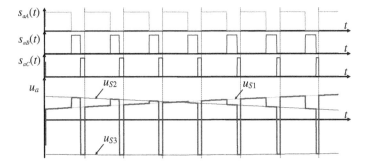

Fig. 2.9 Synthesis of matrix converter output voltages

be expressed as [135]:

$$s_{jA} + s_{jB} + s_{jC} = 1. \tag{2.4}$$

With these restrictions, the three-phase matrix converter has 27 allowed switching states, with 512 (2^9) which are possible [25, 110, 142]. If the load and source voltages are referenced to the supply neutral point "N" then the input/output relationship of voltages and current can be described as follows:

$$
\begin{bmatrix} u_a(t) \\ u_b(t) \\ u_c(t) \end{bmatrix} =
\begin{bmatrix} s_{aA}(t) & s_{aB}(t) & s_{aC}(t) \\ s_{bA}(t) & s_{bB}(t) & s_{bC}(t) \\ s_{cA}(t) & s_{cB}(t) & s_{cC}(t) \end{bmatrix}
\begin{bmatrix} u_A(t) \\ u_B(t) \\ u_C(t) \end{bmatrix} , \tag{2.5}
$$

$$
\begin{bmatrix} i_A(t) \\ i_B(t) \\ i_C(t) \end{bmatrix} =
\begin{bmatrix} s_{aA}(t) & s_{bA}(t) & s_{cA}(t) \\ s_{aB}(t) & s_{bB}(t) & s_{cB}(t) \\ s_{aC}(t) & s_{bC}(t) & s_{cC}(t) \end{bmatrix}
\begin{bmatrix} i_a(t) \\ i_b(t) \\ i_c(t) \end{bmatrix} , \tag{2.6}
$$

$$
\begin{bmatrix} u_{ab}(t) \\ u_{bc}(t) \\ u_{ca}(t) \end{bmatrix} =
\begin{bmatrix} s_{aA}(t) - s_{bA}(t) & s_{aB}(t) - s_{bB}(t) & s_{aC}(t) - s_{bC}(t) \\ s_{bA}(t) - s_{cA}(t) & s_{bB}(t) - s_{cB}(t) & s_{bC}(t) - s_{cC}(t) \\ s_{cA}(t) - s_{aA}(t) & s_{cB}(t) - s_{aB}(t) & s_{cC}(t) - s_{aC}(t) \end{bmatrix}
\begin{bmatrix} u_A(t) \\ u_B(t) \\ u_C(t) \end{bmatrix} .
$$
$$\tag{2.7}$$

The graphical interpretation of MC output voltages formation from pieces of source voltages is presented in Fig. 2.9.

Topologies of bi-directional switches

The three-phase MC topology is constructed using nine bi-directional four-quadrant switches arranged in a matrix, which are capable of conducting currents and blocking voltages of both polarities. There are four main topologies for bi-directional switches, which are shown in Fig. 2.10 [6, 60, 126, 142, 149]. The most simple switch cell is a single-phase diode bridge with an IGBT connected at the centre (Fig. 2.10a). The main advantage of this switch is that only one active device is needed. This approach reduces the cost of the power circuit and the complexity of

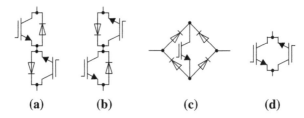

(a)   (b)   (c)   (d)

Fig. 2.10 Bi-directional switch cell configuration: **a** diode bridge with an IGBT configuration, **b** common emitter anti-paralleled IGBT configuration, **c** common emitter anti-paralleled IGBT configuration, **d** anti-paralleled reverse blocking IGBTs (RB-IGBT) configuration

the control. Only one transistor gate drive circuit is needed for each switch cell. The disadvantage is that the conduction losses are relatively high. During the conduction stage, the three devices are conducted (two diodes and IGBT transistor). Moreover, the direction of current through the switch cell cannot be controlled. The two most commonly used configurations of switch cell are named the common emitter anti-paralleled IGBT configuration (Fig. 2.10b) and the common collector anti-paralleled IGBT configuration (Fig. 2.10c). Each of these switch cells consist of two diodes and two IGBT switches that are connected in an anti-parallel arrangement. The diodes are included to provide reverse blocking capability, whereas, the IGBTs enable the independent control of the current direction. Compared to the diode bridge switch cell (Fig. 2.10a), here conduction losses are reduced, because only two devices are conducted in each conduction path. Its disadvantage is the requirement of two gate drive circuits for each IGBTs. For the arrangement shown in Fig. 2.10b, due to its common emitter arrangement, one isolated power supply is required for each bi-directional switch cell. Furthermore, by using common collector bi-directional switch cells (Fig. 2.10c), the number of isolated power supplies required for the gate drive circuits can be reduced to six. Finally, the switch cell is the anti-paralleled reverse blocking IGBTs (RB-IGBT) [130], an arrangement shown in Fig. 2.10d [123, 149]. The main feature of the RB-IGBT is its reverse voltage blocking capability, which eliminates the use of diodes. For this reason there is a reduction in the number of discrete devices and conduction losses. At any instant, there is only one device conducting current in any direction. In this configuration, 18 gate drive circuits and six isolated power supplies is required. Therefore, an anti-paralleled RB-IGBT configuration is generally preferred for creating matrix converter bi-directional switch cells. The element complexity of matrix converters with different switches cells is described in Table 2.3 [21].

Furthermore, other switching devices, besides IGBT, could be used in MCs. If the switching devices used for the bi-directional power switch have a reverse voltage blocking capability, then it is possible to build bi-directional switches. For example, MOS turn-off thyristor (MTOs), GTO thyristor and pure JFET may be an applicable [93].

The first key problem is related to the practical realisation of bi-directional switches. Currently, there are no small bi-directional power switches that are commercially available, so discrete devices need to be used to construct suitable switch cells. These realisations require much more chip area and they produce higher switching losses compared to a completely integrated solution. However, in recent times several different configurations of bi-directional switch cells have appeared on the commercial market. The 1,200 V/200 A IGBT commercially available bi-directional switch cell sample chip, created by Dynex Semiconductor, is shown in Fig. 2.11. This module is a relatively high power device. In the small power semiconductor market this kind of switch module is not available.

From a commercial point of view, it is worth noting that several manufacturers have already produced integrated power modules for MC. The traditional solution tends to concentrate on a single power module, with the switches corresponding to one leg of the converter. The prototype of one-phase leg of an MC with RB-IGBTs is shown in Fig. 2.12 [123, 149]. The power modules included six 600 V/100 A RB-IGBTs, connected as shown in Fig. 2.13.

However, it is also possible to find modules containing the whole power stage of the MC. This arrangement leads to a very compact converter with the potential for substantial improvements in efficiency. The first three-phase-to-three-phase matrix switch power module was built by Eupec in co-operation with Siemens, using transistors connected in the common collector configuration [60, 122]. It contains all 18 necessary IGBTs and diodes of the 3 × 3 switch matrix in a single housing (Fig. 2.14). This module is named EconoMAC [60]. This type of packaging will have important benefits in terms of circuit losses. The stray inductance and connect resistances can be minimised.

Table 2.3 The MC element complexity with different switch cells

Switch cell configuration	Transistors	Diodes	Isolated power supply	Gate drive circuits
Figure 2.10a	9	36	9	9
Figure 2.10b	18	9	9	18
Figure 2.10c	18	9	6	18
Figure 2.10d	18	0	6	18

(a)   **(b)**

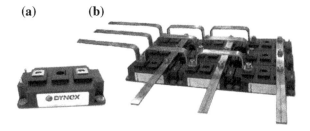

Fig. 2.11 Chip of commercially available bi-directional switch cell: **a** single switch, **b** matrix-connected nine switches

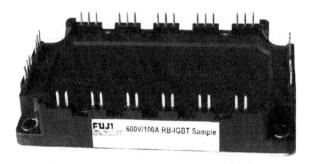

Fig. 2.12 Prototype of 600 V/100 A RB-IGBT module—photograph [123]

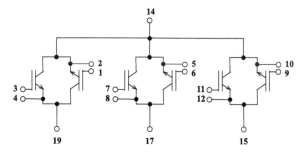

Fig. 2.13 Prototype of 600 V/100 A RB-IGBT module—topology structure [123]

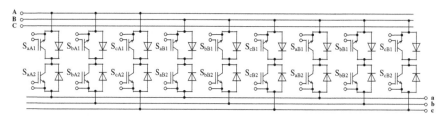

Fig. 2.14 Power stage configuration of EconoMAC module [60]

The all-in-one MC configuration prototype module with RB-IGBTs was introduced by FUJI Electric in 2011. The prototype 1,200 V/50 A module is shown in
Fig. 2.15, whereas the internal structure is depicted in Fig. 2.16 [100].

Nowadays, several bi-directional switches, one-phase leg matrix converter or
three-phase matrix configuration power switch modules are proposed by various
companies [25, 26, 60, 100, 122, 123, 149]. Based on the above review it can be
said that the number of power semiconductors for matrix converter application will
systematically increase over time. The market for these devices depends on the development of MC technology. The main focus of development is the reduction of costs,
size and increase in reliability [6, 98, 99].

Fig. 2.15 Photography of RB-IGBT matrix configuration module

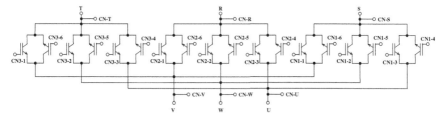

Fig. 2.16 Power stage configuration of RB-IGBT matrix configuration module

Commutation strategies

One of the main issues in control of the MC is the current commutation [25, 142]. The switches used in the MC are not protected by the DC-link capacitor, which is typical of the classical VSI, since there are no natural freewheeling paths. The current commutation between switches in the MC is more difficult to achieve than the VSI [105]. When considering commutation strategies for matrix converters two general rules must be adhered [29, 135]:

1. commutation should not cause a short circuit between the two input phases, because the consequent high circulating current might destroy the switches;
2. commutation should not cause an interruption of the output current because the consequent overvoltage might likely destroy the switches.

The switches have to be capable of being turned on and turned off in such a way as to avoid short circuits and sudden current interruptions. The commutation has to be actively controlled at all times with respect to the two above-mentioned basic rules. In order to explain the strategy it is helpful to refer to the simplified commutation circuit shown in Fig. 2.17 [142]. Taking into consideration the basic rules, it is important that no two bi-directional switches are switched on at any given instant (Fig. 2.17a). When the switches are turned on simultaneously, then the voltage sources will be shorted directly and the switches will be damaged due to over-currents. In the case where all the switches are turned off simultaneously (Fig. 2.17b), in the first instant after the switching-off an over-voltage will be generated which could destroy the semiconductors. The spikes of over-voltage depend on load current and duration of current interruption ($u_{\text{spikes}} = L di_L/dt$). These two considerations

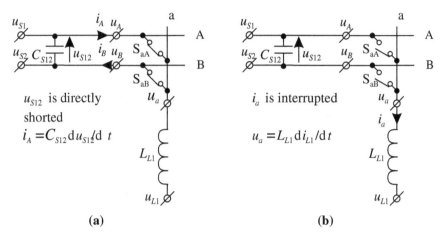

Fig. 2.17 Disallowed switch configurations in MC: **a** short circuit of capacitive input, **b** open circuit of inductive load

cause a conflict since semiconductor devices cannot be switched instantaneously between states because of propagation delays and finite switching times.

Various methods have been proposed to avoid this difficulty and to ensure successful commutation. To fulfil commutation requirements some knowledge of the commutation conditions is mandatory, e.g. the polarity of input voltage between the involved bi-directional switches or the polarity of load current. This part of the chapter offers an overview of different current commutation strategies in the MC. The first and second presented commutation methods intentionally break the rules of MC current commutation, which are mentioned above, and need extra circuitry to avoid destruction of the switches.

The first is based on a dead-time method which is commonly used in inverter systems. Using dead-time commutation would cause an open circuit of the load [32, 142]. This would result in large voltage spikes across the switches. This necessitates the use of snubbers or clamping devices across the switch cells to provide a path for the load current. In this method the commutation losses are relatively high. All commutation energy is lost in snubbers or clamping devices. Furthermore, the clamping devices increase converter volume. The MC snubbers are more complicated than snubbers in the VSI, due the bi-directional nature of the switch cells.

The second current commutation is known as the overlap commutation method [32]. This method also breaks the rules of MC current commutation. In overlap current commutation, the incoming switch is turned on before the outgoing switch is turned off [13]. During the overlap period extra line inductance is added to slow the rise of the current. The inductors are in the main conduction path, and the conduction losses will be increased. Furthermore, during the overlap period the load voltages are deformed. The switching time for each commutation is increased and will vary

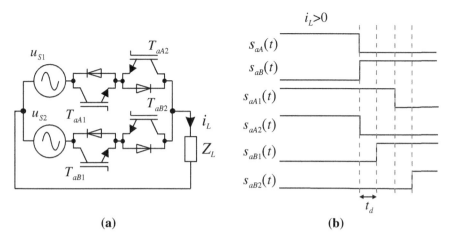

(a)   **(b)**

Fig. 2.18 For step commutation of bi-directional switches based on current direction for $i_L > 0$: **a** general commutation circuit of two bi-directional switches, **b** timing diagram

with level of voltage and inductor value. As a consequence, the switching frequency is decreased.

These two current commutation methods have disadvantages. Therefore, it is preferable to use advanced commutation methods. The commutation problem has been solved with the development of several multistep commutation strategies that allow safe operation of the switches. The most common solution is the four-step commutation strategy (or semi-soft current commutation) introduced by Burany in 1989 [20]. In this method the direction of current flow through the commutation cells can be controlled. In order to explain the strategy it is helpful to refer to the simplified commutation circuit shown in Fig. 2.18. The strategy assumes that when the output phase is connected to an input phase, both the IGBTs of the bi-directional switch S_1 have to be turned on simultaneously. The following example assumes that the load current ($i_L > 0$) is in the direction as shown in Fig. 2.18a and the upper bi-directional switch (S_1) is closed. In this method, the current direction is used to determine which device in the active switch cell is not current conducting. The commutation process is shown as a timing diagram in Fig. 2.18b. In the beginning, both IGBTs of switch S_1 are turned on in the same instant. In the first step, the IGBT T_{aA2}, which is not conducting the load current, is turned off. In the second step, after delay interval time t_d, the transistor T_{aB1} that will conduct the current is turned on. This allows both cells to be turned on without short circuiting the input phases and provides a path for the load current. Depending on the instantaneous input voltages, there are two kinds of commutation process after the second step. If $u_{S2} > u_{S1}$ and $i_L > 0$, then the conducting diode of switch cell S_1 could be reverse biased and a natural commutation could take place. In the third step the IGBT T_{aA1} is turned off. If there is no natural commutation during the second step, then a hard commutation happens when, in the third step, IGBT T_{aB1} is turned off. A short time later, in the fourth

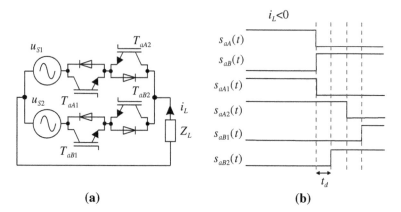

(a)   **(b)**

Fig. 2.19 For step commutation of bi-directional switches based on current direction for $i_L < 0$: **a** general commutation circuit of two bi-directional switches, **b** timing diagram

Fig. 2.20 Four-step commutation based on current direction switching diagram for two bi-directional switches from Fig. 2.18a ($i_L > 0$) and from Fig. 2.19a ($i_L < 0$)

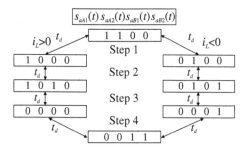

step, transistor T_{aB2} is turned on to also allow the conduction of negative currents. The time delay has to be set to a value higher than the maximum propagation time of the IGBT signals. In this commutation method half of the commutation process is soft switching and half is hard switching. As a result this method is often called semi-soft current commutation [142]. Problems occur, however, at low current levels when the direction of the current is not certain and incorrect decisions are made as to which switches conduct the load current. This can be a problem if no protection device is employed. The simplified commutation circuit and timing diagram for the second condition $i_L < 0$ is shown in Fig. 2.19. A state diagram of the commutation process for a four-step commutation sequence between two bi-directional switches from Figs. 2.18 and 2.19a is shown in Fig. 2.20.

A simplification of the four-step current commutation method is to only gate the conducting device in the active switch cell [105, 124]. The non-conducting IGBTs are turned off during the commutation process and steady-state condition. Then there is created a simple two-step current commutation strategy, as shown in Fig. 2.21a. Current reversal is achieved by turning on the reverse transistor in the switch cell when the current falls below a threshold level (Fig. 2.21b). If the current achieves a value above the threshold level in the opposite direction, the initial device is turned

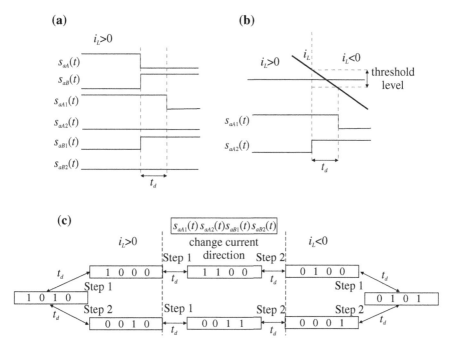

Fig. 2.21 Simple two-step current commutation: **a** timing diagram for commutation between two switches, **b** current reversal using threshold detection, **c** state diagram for the commutation process

off. A state diagram for the commutation process for a simple two-step commutation sequence between two bi-directional switches from Fig. 2.18a is shown in Fig. 2.21c.

This commutation method has practical limitations. During the current reversal period the current direction is unknown, because the current reversal switch is subject to hysteresis. During this time period the commutation cannot take place. Since the direction of current is unknown, the correct device that will conduct the current cannot be determined. The second disadvantage is that current direction can be difficult to determine, especially in high power drives when the levels of current are low and when the load current has to be within a threshold level. Then the threshold level may also be relatively large. This may result in a distorted current waveform.

For current commutation techniques it is required to know the load phase current direction. High precision determination of the direction of current is a key issue. Any inaccuracies cause errors, which result in switching losses and the possibility of destroying switch cells. To solve this problem a new technique has been developed [33]. This technique uses the voltage across the bi-directional switch to determine the current direction. This conception is based on an intelligent gate drive circuit. In addition, to control the IGBT this gate driver can also detect the current direction and enables the exchange of information between other gate driver devices. This process ensures that all gate drivers can operate with safe commutation.

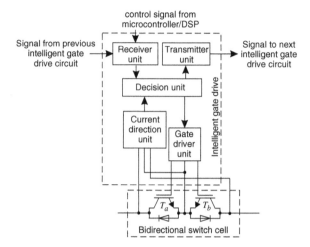

Fig. 2.22 Simplified block diagram of intelligent gate driver

Fig. 2.23 Voltage measurements in switch cell

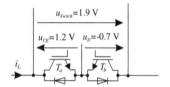

A simplified block diagram of the intelligent gate driver is shown in Fig. 2.22 [33]. Current direction detection is based on voltage measurements across each of the devices in the commutation cell (Fig. 2.23). In the case which is shown in Fig. 2.23 the voltages are defined as follows: $u_{CE} = 1.2\,\text{V}$ and depends on transistors used, $u_D = -0.7\,\text{V}$. When load current is in the opposite direction the reverse situation exists. The current polarity is detected and can be calculated on the basis of the measured results. Information about current direction is sent to all intelligent gate drivers on the same output line.

The commutation process is as follows [33, 140]. Taking into consideration the current direction as shown in Fig. 2.18 and with the switch cell S_1 conducting (T_{aA1} is turned on), then the current direction information from the cell S_1 gate drive is passed to the gate drive for cell S_2. Transistor T_{aB1} is turned on and after delay time the transistor T_{aA1} is turned off. After a short time interval the current direction information is taken from the detection circuit in switch cell "B" rather than switch cell "A". The commutation is now complete. A timing diagram and state diagram for the commutation process for a two-step commutation sequence between two bidirectional switches with intelligent gate driver is shown in Fig. 2.24a, b, respectively. In this commutation method the current direction is known at any instant. If the detection circuit determines that the load current has fallen to zero then the intelligent gate driver sends this information to another gate driver.

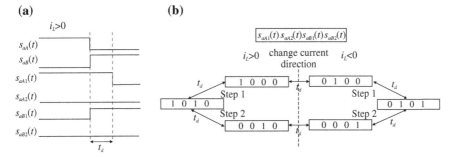

Fig. 2.24 Two-step current commutation with intelligent gate driver: **a** timing diagram for commutation between two switches, **b** state diagram for the commutation process

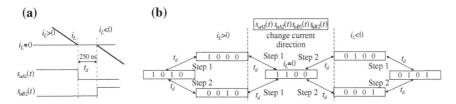

Fig. 2.25 Two-step current commutation with intelligent gate driver including propagation delay compensation: **a** timing diagram for commutation between two switches, **b** state diagram for the commutation process

A two-step current commutation method with intelligent gate driver also has special cases, where a potential difficulty occurs. One such case is when a commutation between switch cells occurs and the load current changes direction. The problem is due to the propagation delay in sending the data on the current direction to the next gate drive switch cell [33]. If a commutation between switch cells occurs first and then the information about the current direction reaches the next gate driver, then the wrong switch is turned on. This problem is solved by having a short dead time when the current reaches zero. During this time period no switches are turned on, as shown in Fig. 2.25. The reverse device is not gated until the new information is received by the other gate drivers. This delay time is small and only depends on the propagation delay inherent in the communication lines (typically 250 ns). This small dead time does not unduly distort the load current waveforms. A state diagram for the commutation process for a two-step commutation sequence between two bi-directional switches with intelligent gate driver and propagation delay compensation is shown in Fig. 2.25 [32, 33].

Some commutation was based on input voltage polarity measurements. To introduce this method input line-to-line voltages have to be measured in order to detect the polarity of the voltage across the two bi-directional switches involved in the commutation process. The operating principle is to provide, by proper control of active devices, the output current with freewheeling paths. Similar to the commutation method based on current direction there are different number of commutation steps. The first is a four-step commutation strategy, which was presented in [3]. In general, the switching sequence depends on the voltage level switches involved in the commutation process. In this strategy both IGBTs of the conducting bi-directional switch are turned on. When commutation between switch cells occurs, the first stage is to determine the voltage level at the turned on and the turned off switch cells. This is needed to identify within the two commutating bi-directional switches the active devices that will operate as freewheeling devices. In general, the freewheeling devices are as follows:

1. the devices which allow the current flow from source to load in the lower input voltage phase;
2. the devices which allow the current flow from load to source in the higher input voltage phase.

After determination of freewheeling devices, the second action is switch commutation in the following four steps:

Step 1: the freewheeling device of the incoming switch is turned on;
Step 2: the non-freewheeling device of the outgoing switch is turned off;
Step 3: the non-freewheeling devices of the incoming switch is turned on;
Step 4: the freewheeling device of the outgoing switch is turned off.

The above described commutation process is depicted in Fig. 2.26a. A state diagram of the commutation process for a four-step commutation sequence between two bi-directional switches with voltage polarity measurement is shown in Fig. 2.26b [29].

Another input voltage measurement-based commutation strategy was presented by Ziegler and Hofmann in [152]. It is based on the basic operating principle of providing a freewheeling path for both output current polarities at any given time, for devices with either steady- or transient-state combinations. This commutation method is called METZI and is based on the detection of the six time intervals as

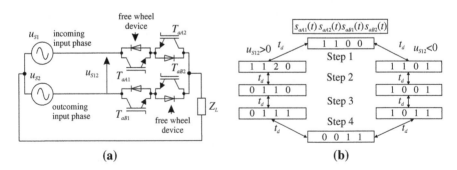

(a)   (b)

Fig. 2.26 Four-step voltage polarity measurement commutation method; **a** general commutation circuit, **b** switching diagram

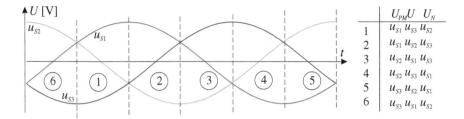

Fig. 2.27 Description of input voltages intervals in METZI method

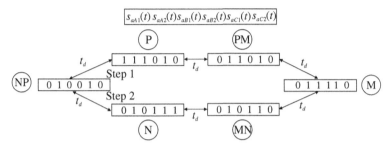

Fig. 2.28 Switching diagram for two-step METZI commutation method

shown in Fig. 2.27. METZI commutation is considered for each load phase. The two-step commutation is obtained by using more active devices in either steady or transient states. Four of the six switches are turned on in every major state. Two switches are turned on to ensure a bi-directional path for the load current, and the redundant two switches are turned on for two-step commutation. In every time interval (Fig. 2.27) one input line has the highest voltage U_P, one the lowest U_N and one the middle U_M. There are six switching states, three major states (P, M, N) and three intermediate states (PM, MN, NP) for each output phase, as shown in the state diagram in Fig. 2.28. Then two-step commutation rules are defined as follows [58]:

Step 1: turn off all switches which will not be switched on in the target base state; the auxiliary state will be reached;

Step 2: turn on the switches of the target base state; the target state will be reached.

Figure 2.29 shows the example of commutation from phase P to phase M with METZI commutation method.

The implementation of this method requires a very accurate measurement of the input voltage. When implementing this method, some problems are presented at the zero crossing points of the line-to-line voltage, as depicted in Fig. 2.30. The improved commutation method, which takes into account critical regions during the zero crossing point of the line to line voltage, is presented in [150]. The critical sequence is replaced by two uncritical sequences which will be commutated to the

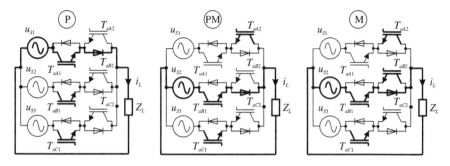

Fig. 2.29 Two-step METZI commutation from phase P to phase M

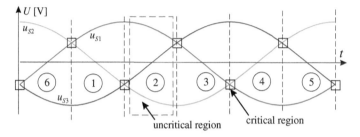

Fig. 2.30 Critical regions for commutation strategies based on the voltage polarity detection

remaining third input phase and then to the desired destination phase. A few other solutions of commutation in critical regions are presented in papers [53, 96, 97, 118].

The disadvantages of the previously presented commutation strategies can be partially avoided by using a commutation strategy where both output current and input line-to-line voltage sign are measured. This strategy was proposed in [29] and is realised by three steps. The first key advantage of this commutation strategy is that output current commutates between the off-going and on-going bi-directional switch always at the same instant with respect to the beginning of the commutation process. The second advantage of the proposed three-step commutation strategy is to decrease the value of the minimum duty cycle the converter is able to apply.

The strategy assumes that when the output phase is connected to an input phase both the IGBTs of the bi-directional switch S_1 have to be turned on simultaneously, allowing an automatic output current reversal. With the knowledge of the output current direction and line-to-line input voltage polarity between off-going and on-going phases the commutation rules are defined. There are two different three-step switching sequences, which depend on the voltage polarity. If the output current is positive, then the following two-switching sequences are used, where u_{S12} is denoted as in Fig. 2.26 [29]:

Sequence 1 $u_{S12} > 0$: In the first step the IGBT that is not carrying the current in the off-going switch cell is turned off and simultaneously the IGBTs that will carry the current in the on-going switch cell is turned on. Then, in the second step, the IGBT

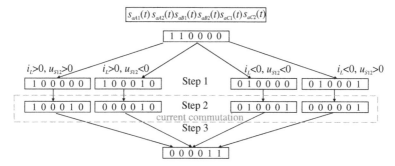

Fig. 2.31 Switching diagram for both three-step voltage and current polarity measurement commutation method

still gated in the off-going switch cell is turned off. In the last step the IGBTs that will not carry the current in the on-going switch-cell are turned on.

Sequence 2 $u_{S12} < 0$: In the first step the IGBT that is not carrying the current in the off-going switch cell is turned off first. In the second step, the IGBT that will carry the current in the on-going switch cell is turned on. In the last step the IGBT still gated in the off-going switch cell is turned off and simultaneously the IGBT that will not carry the current in the on-going switch cell is turned on.

A state diagram for the commutation process for a three-step commutation sequence between two bi-directional switches with both voltage and current polarity measurement is shown in Fig. 2.31. The dashed contour indicates the state of the switching sequences in which the current commutates. In Fig. 2.31 the commutation sequence for negative polarity of output current is also shown. Detailed information about the advantages of a three-step commutation strategy is presented in [29].

The commutation strategies presented above are the most well known and concern the bi-directional switch cell with IGBT and reverse diodes. In the literature can be found more variants of the presented commutation with different numbers of steps [56, 58, 118, 119, 137, 139, 151].

Recently, a reverse-blocking insulated gate bipolar transistor (RB-IGBT) was developed as an alternative solution for MC bidirectional switches [67]. This RB-IGBT is based on the ultrathin-wafer technology, and one unique feature is that its reverse leakage current is closely related to u_{GE}. When the RB-IGBT operates in reverse blocking condition, then a positively biased u_{GE} can reduce the reverse leakage current significantly. In an MC with the anti-parallel RB-IGBT, the previously presented commutation methods based on load current direction measurements or input voltages polarity measurements can be used, as presented in [67, 71]. However, a novel commutation method for an MC with RB-IGBT has been developed. A detailed description of this method is shown in [123], and it is based on RB-IGBT properties in reverse blocking condition. This method is implemented by utilizing load current direction signals and input voltage relations. A simplified commutation circuit of RB-IGBT is shown in Fig. 2.32. Taking into account the circuit in Fig. 2.32,

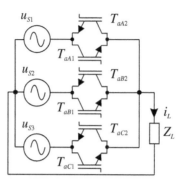

Fig. 2.32 Simplified commutation circuit with RB-IGBT

for different combinations of load current direction, and input voltage relationship, there are eight possibilities altogether, as summarised in Table 2.4 [123].

Soft switching is the commutation method which removes the switching losses associated with hard switching, allowing for higher switching frequencies to be used with reduced EMC emissions. Soft switching is the switching of devices when either a zero voltage or zero current condition occurs. The theory of soft commutation in MC is well known. The techniques developed fall into two categories: resonant switch circuits [32, 107, 136] and auxiliary resonant circuits [14, 18, 30, 57, 131].

The example of a resonant switch cell with common emitter connected IGBT and diode bridge arrangement is shown in Fig. 2.33. The soft switched cell shown in Fig. 2.33a consists of a standard common emitter anti-parallel IGBT and diode cell with one extra IGBT, two diodes, a capacitor, inductor and voltage source [32, 136]. For the following explanation, a simplified commutation circuit is connected similar to that in Fig. 2.26, but using a soft switch cell (cell one and cell two are assumed to be the incoming and outgoing switches, respectively). All transistors are turned on and turned off simultaneously. In the commutation between incoming and outgoing switches current flows from the supply of cell one through C_R and also through D_3, L_R, E, T_A and D_2. Current also flows through C_R of cell two and capacitor voltage u_{CR} charges linearly until it equals E. The element L_R and C_R of both cells forms a resonant circuit. When u_{CR} is equal to zero, D_1 starts to conduct ($I_{D1} = I_{LR} - I_L$). and inductor L_R discharges linearly through D_1 and D_3. When $I_{LR} = I_L$ transistor T_1 starts to conduct ($I_{T1} = I_L - I_{LR}$), and the inductor current is still linearly discharging. When $I_{LR} = 0$, transistor T_2 conducts the full i_L. This ensures that the main switches switch under zero voltage conditions and that the auxiliary switch switches under zero current conditions. The major problem with this soft switch is that the voltage source E is difficult to realise in a practical system.

It is seen from Fig. 2.33b that the presented bi-directional switch consists of a bridge rectifier (D_1, D_2, D_3, D_4), two transistors (T_1, T_2) a small inductor (L_R) and a capacitor (C_R). The transistors are turned on and turned off simultaneously. Current direction of inductor L_R and voltage polarity of capacitor C_R are unidirectional. The

Table 2.4 Commutation method of RB-IGBT

Condition $u_{S1} > u_{S2}$ and $u_{S1} > u_{S3}$ and $i_L > 0$	T_{aA1}	T_{aA2}	T_{aB1}	T_{aB2}	T_{aC1}	T_{aC2}
Initial state:	1	1i	0	1	0	1
Step 1:	1	0	0	1i	0	0
Step 2:	1	0	1	1i	0	0

Condition $u_{S1} < u_{S2}$ and $u_{S1} > u_{S3}$ and $i_L > 0$	T_{aA1}	T_{aA2}	T_{aB1}	T_{aB2}	T_{aC1}	T_{aC2}
Initial state:	1	1i	1	0	0	1
Step 1:	0	1i	1	0	0	1
Step 2:	0	1	1	1i	0	1

Condition $u_{S1} > u_{S2}$ and $u_{S1} < u_{S3}$ and $i_L > 0$	T_{aA1}	T_{aA2}	T_{aB1}	T_{aB2}	T_{aC1}	T_{aC2}
Initial state:	1	1i	0	1	1	0
Step 1:	1	0	0	1i	1	0
Step 2:	1	0	1	1i	1	0

Condition $u_{S1} < u_{S2}$ and $u_{S1} < u_{S3}$ and $i_L > 0$	T_{aA1}	T_{aA2}	T_{aB1}	T_{aB2}	T_{aC1}	T_{aC2}
Initial state:	1	1i	1	0	1	0
Step 1:	0	1i	1	0	1	0
Step 2:	0	1	1	1i	1	0

Condition $u_{S1} > u_{S2}$ and $u_{S1} > u_{S3}$ and $i_L < 0$	T_{aA1}	T_{aA2}	T_{aB1}	T_{aB2}	T_{aC1}	T_{aC2}
Initial state:	1i	1	0	1	0	1
Step 1:	1i	0	0	1	0	0
Step 2:	1	0	1i	1	0	0

Condition $u_{S1} < u_{S2}$ and $u_{S1} > u_{S3}$ and $i_L < 0$	T_{aA1}	T_{aA2}	T_{aB1}	T_{aB2}	T_{aC1}	T_{aC2}
Initial state:	1i	1	1	0	0	1
Step 1:	0	1	1i	0	0	1
Step 2:	0	1	1i	1	0	1

Condition $u_{S1} > u_{S2}$ and $u_{S1} < u_{S3}$ and $i_L < 0$	T_{aA1}	T_{aA2}	T_{aB1}	T_{aB2}	T_{aC1}	T_{aC2}
Initial state:	1i	1	0	1	1	0
Step 1:	1i	0	0	1	1	0
Step 2:	1	0	1i	1	1	0

Condition $u_{S1} < u_{S2}$ and $u_{S1} < u_{S3}$ and $i_L < 0$	T_{aA1}	T_{aA2}	T_{aB1}	T_{aB2}	T_{aC1}	T_{aC2}
Initial state:	1i	1	1	0	1	0
Step 1:	0	1	1i	0	1	0
Step 2:	0	1	1i	1	1	0

where *1* RB-IGBT turn-on and non-conducting current, *1i* RB-IGBT turn-on and conducting current, *0* RB-IGBT turn-off

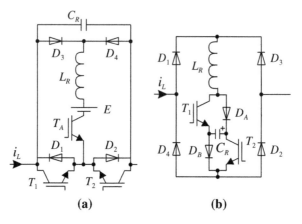

Fig. 2.33 Soft switched cell with: **a** common emitter connected IGBT, **b** diode bridge arrangement

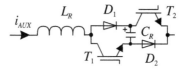

Fig. 2.34 Unidirectional auxiliary resonant components

circuit operation from Fig. 2.33b is described in [32, 107]. The capacitor voltage polarity before turning on is shown as in Fig. 2.33b, to reverse biasing diodes D_A and D_B. When T_1 and T_2 are turned on, i_{T1} and i_{T2} will increase from zero resulting in zero current switching. The voltage of C_R decays to zero and the diodes D_A and D_B become forward biased. Then the current flows through both IGBT and its series diode D_A and D_B. Next, the turning off process is as follows. When T_1 and T_2 are turned off, the switch voltages (U_{T1} and U_{T2}) will increase from zero resulting in zero voltage switching. Hence, the capacitor serves as a snubber to eliminate the voltage spike. The inductor current will fall to zero and the capacitor will be charged ready for the another turning on process.

In the solution with the presented soft switch cells the switching losses are decreased but the conduction losses are increased due to the extra devices in the main conduction path. Another disadvantage of this switch configuration is the increase in the number of components.

The second soft commutation idea is to use auxiliary resonant components on each output phase of the matrix converter in the attempt to force the current or voltage to zero during commutation [18]. The proposed circuit using auxiliary resonant switches is shown in Fig. 2.34 [18]. Two such switches are used for each output phase, one for positive load current and one for negative load current as shown in Fig. 2.35.

The circuit operation of Fig. 2.34 is similar to that of the circuit in Fig. 2.33b. The auxiliary resonant switch in the configuration shown in Fig. 2.35 is to conduct the

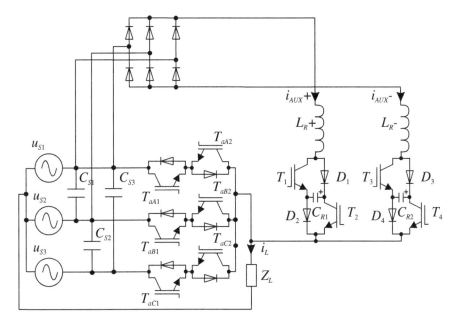

Fig. 2.35 Single phase of MC with the auxiliary resonant components

load current during commutation allowing ZCS of the main switches. Because of this, the conduction loss will not be significantly increased compared to a converter using non-resonant techniques. Auxiliary resonant circuits will also increase the component count but not to the same extent as circuits using soft switching cells. An advantage of this type of circuit is that the auxiliary resonant components can be disabled when operating at low voltage or current. Another concept of auxiliary resonant components on each output phase of the MC is presented in [57].

Another concept of soft switching with auxiliary resonant components is the auxiliary resonant commutated pole structure (ARCP), which has been fully analysed in [14, 30, 131]. Three different ARCP-MC topologies previously proposed are shown in Fig. 2.36. Unfortunately, the control complexity of the ARCP-MC is significantly higher than that of the MC. The number of elements is also increased.

All these soft switching circuits (Figs. 2.33, 2.34, 2.35, Fig. 2.36) significantly increase the component count in the MC, and increase conduction losses. Furthermore, a modification of the MC control algorithm is required.

Protection Issues

In the previous subsection a convenient manner of commutating the IGBTs was discussed. It was noted that in some cases there are over-voltages that should be managed appropriately to avoid semiconductor destruction [95, 105]. Other sources of over-voltages are grid perturbations and fault states in the load and, therefore, it is important to have a method of dealing with these phenomena. An effective and

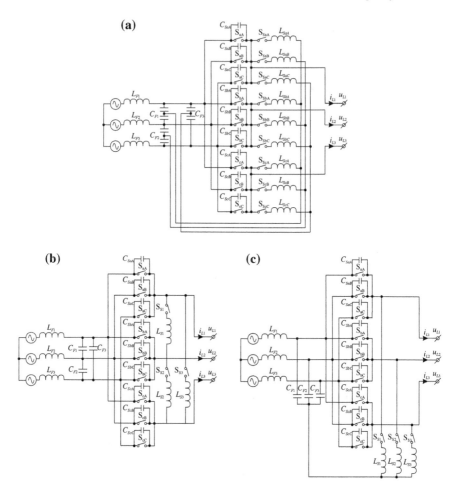

Fig. 2.36 MC auxiliary resonant commutated pole (MC-ARCM), **a** concept I, **b** concept II, **c** concept III [131]

robust protection scheme is an important element in the implementation of a stable and reliable power stage in MCs.

In [135] the first protection circuit was proposed, consisting of input and output diode bridges, an electrolytic capacitor and its charge and discharge circuit. Figure 2.37 shows this over-voltage circuit [94, 95], which is the most common solution for avoiding over-voltages coming from the grid and from the motor. This clamp configuration uses 12 fast-recovery diodes to connect the capacitor to the input and output terminals. Then a capacitor takes the commutation energy and the resistor can discharge the capacitor. When over-voltage occurs, (in the case of a hard commutation and abnormal operation of the motor) the diode conducts and the *RC* circuit maintains the voltage level at a safe value. In normal operation, the diodes

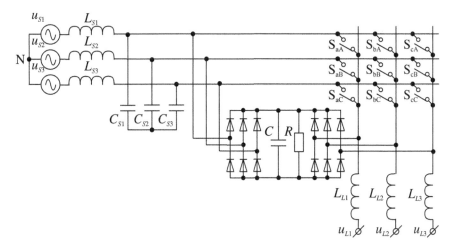

Fig. 2.37 Matrix converter with a 12-diode protected clamp circuit

are off and the clamp circuit has no influence on the MC operation. In the case of a drive system, when all the switches are turned off, the current in the load is suddenly interrupted. The energy stored in the motor leakage inductance has to be discharged without creating dangerous over-voltages. Then, a shut-down of the converter can be made by reducing the power to the machine, causing no interruption of the motor current.

The energy stored in DC capacitors is discharged in the resistor [5, 6, 75] or is used to feed the control electronics and to magnetise the motor-ride-through capability [74]. This over-voltage protection circuit has the advantages of being very simple, it has low hardware requirements and simple control strategies. However, this circuit (Fig. 2.37) has some drawbacks, such as the high number of required semiconductor devices (12 fast-recovery auxiliary diodes). The reduction of diodes to six is possible in the over-voltage protection circuit, depicted in Fig. 2.38. In this topology six additional diodes for the power bi-directional switches are used [101]. In both these protected circuits the electrolytic capacitor has a large volume which constrains the lifetime of the system. The discharge circuit by a DC chopper increases the number of power devices.

On the other hand, a varistor protection and a suppressor diode protection were proposed in [95]. Figure 2.39 shows the varistor location. Varistors are connected at the input and at the output terminals of MC. These protections are very useful for a small capacity system, but not suitable for a large capacity system. The protection strategy with varistor over-voltage protection allows the removal of the large and expensive diode clamp. The input varistor has to protect the converter switches from the voltage surges coming from the AC mains. At the output side, the varistor protect the MC power stage devices from a hard converter shut-down or a converter error during a commutation process. During normal operations the losses caused

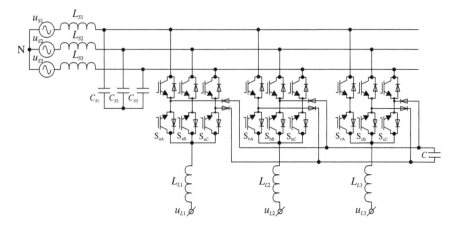

Fig. 2.38 Matrix converter with a 6-diode protected clamp circuit

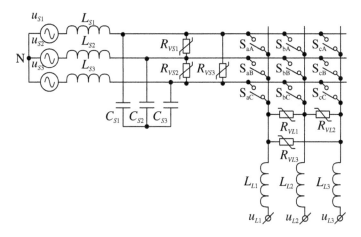

Fig. 2.39 Matrix converter with varistor protection

by the varistors are not worth mentioning. Unfortunately, the varistor triangles by themselves are not sufficient to guarantee during a converter shut-down a reliable protection of the IGBTs. Then a simple extra circuit to protect each IGBT is required. A problem occurs when a turning off bi-directional switch achieves its blocking capability with a certain delay in respect to the others. The neighbouring IGBT having its full blocking capability may get the maximum clamping voltage of the varistor, causing damage to this device. In order to protect the single IGBT, a circuit made up with a suppressor diode is added to any IGBTs. Figure 2.40 shows the IGBT with a suppressor diode. To ensure a good performance and lifetime of the MC a combination of both varistor and suppressor diode protection is required [95].

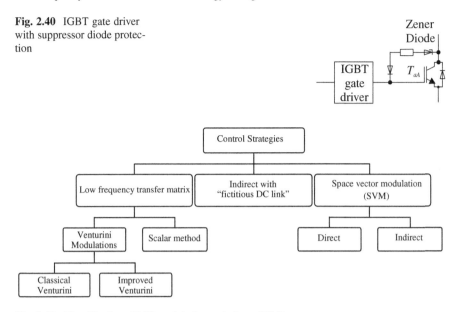

Fig. 2.40 IGBT gate driver with suppressor diode protection

Fig. 2.41 Classification of MC modulation techniques [126]

As shown in Fig. 2.40, the protection IGBT circuit with suppressor diode uses the Zener diode, with a high breakdown voltage [95]. But the Zener breakdown voltage has to be lower than the maximum blocking voltage of the IGBT. Then, the operation of the suppressor diode is as follows: if the collector–emitter voltage of the IGBT increases to a value higher than that of the breakdown voltage of the suppressor diode, this diode becomes conductive. The gate of the IGBT is charged again. Each IGBT can be protected in this way because the IGBT becomes conductive and destructive voltage is eliminated.

Modulation techniques

The complexity of the matrix converter topology makes the study and the determination of suitable modulation strategies a hard task. A review of the well-known modulation techniques is presented in this paragraph. From this unitary point of view, some modulation techniques are described and compared with reference to maximum voltage transfer ratio. Several modulation strategies have been proposed in previous work [3, 4, 17, 21–28, 54, 55, 61, 62, 65, 106, 108–113, 133–135, 142, 148, 155, 156]. These modulation strategies give different voltage conversion ratios and the number of commutations employed in each modulation strategy is different. A modulation strategy can be broadly classified into three categories, depending upon the type of calculation of switch states. Figure 2.41 shows the tree of such classification of MC modulation techniques [126].

The MC bi-directional power switches work with a high switching frequency. A low frequency load voltage of variable amplitude and frequency can be generated by modulating the duty cycle of the switches using their respective switching functions

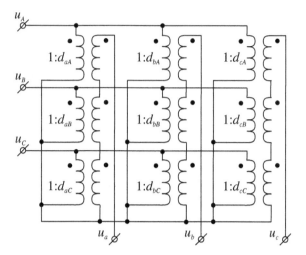

Fig. 2.42 Matrix converter averaged-switching-cycle representation

s_{jK}. A modulation duty cycle should be defined for each switch in order to determine the average behaviour of the MC output voltage waveform [110, 135, 142]. The modulation duty cycle is defined by:

$$d_{jK}(t) = \frac{t_{jK}}{T_{\text{Seq}}}, \tag{2.8}$$

where t_{jK} represents the time when switch S_{jK} is turned on and T_{Seq} represents the time of the complete sequence in the PWM pattern, and $0 < d_{jK} < 1$. Based on the switch duty-ratios, the averaged output voltages and the averaged input currents can be related to the input voltages and the output currents, respectively, as [142]:

$$\bar{\mathbf{u}}_L = \mathbf{M}(t)\mathbf{u}_S, \qquad \bar{\mathbf{i}}_S = \mathbf{M}^T(t)\bar{\mathbf{i}}_L, \tag{2.9}$$

where:

$$\mathbf{M}(t) = \begin{bmatrix} d_{aA}(t) \; d_{aB}(t) \; d_{aC}(t) \\ d_{bA}(t) \; d_{bB}(t) \; d_{bC}(t) \\ d_{cA}(t) \; d_{cB}(t) \; d_{cC}(t) \end{bmatrix}. \tag{2.10}$$

The matrix $\mathbf{M}(t)$ is known as the modulation matrix or low-frequency transfer matrix. Based on these relationships in (2.9) and (2.10), a matrix converter on a switching-cycle averaged basis can be represented by nine ideal transformers with varying turn-ratios, as shown in Fig. 2.42 [115].

Venturini Modulation Techniques

In 1980, Venturini and Alesina presented a PWM modulation method for the control of MCs [135]. The proposed method by these authors is known as the classical Venturini modulation or the direct transfer function approach. The modulation problem assumes that a set of sinusoidal load voltages, $u_L = [u_{L1}(t), u_{L2}(t), u_{L3}(t)]^T$ and source currents, $i_S = [i_{S1}(t), i_{S2}(t), i_{S3}(t)]^T$ are required:

$$\mathbf{u}_L = qU_{Lm} \begin{bmatrix} \cos(\omega_L t) \\ \cos(\omega_L t - 120) \\ \cos(\omega_L t + 120) \end{bmatrix}, \quad \mathbf{i}_S = qI_{Sm} \begin{bmatrix} \cos(\omega t + \varphi) \\ \cos(\omega t - 120 + \varphi) \\ \cos(\omega t + 120 + \varphi) \end{bmatrix}. \quad (2.11)$$

A set of input voltages and an assumed set of output currents are described as follows:

$$\mathbf{u}_S = U_{Sm} \begin{bmatrix} \cos(\omega_L t) \\ \cos(\omega_L t - 120) \\ \cos(\omega_L t + 120) \end{bmatrix}, \quad \mathbf{i}_L = I_{Lm} \begin{bmatrix} \cos(\omega t + \varphi) \\ \cos(\omega t - 120 + \varphi) \\ \cos(\omega t + 120 + \varphi) \end{bmatrix}, \quad (2.12)$$

where: q is the voltage transfer ratio, ω and ω_L are the input and output pulsation, respectively, and φ_S and φ_L are the input and output phase displacement angles, respectively. The low-frequency transfer matrix proposed by Venturini is described in [135] as:

$$\mathbf{M}(t) = \mathbf{M}^+(t) + \mathbf{M}^-(t), \quad (2.13)$$

where:

$$\mathbf{M}^+(t) = \frac{\alpha_1}{3} \begin{bmatrix} 1 + 2qm^+(0) & 1 + 2qm^+(-\frac{2\pi}{3}) & 1 + 2qm^+(-\frac{4\pi}{3}) \\ 1 + 2qm^+(-\frac{4\pi}{3}) & 1 + 2qm^+(0) & 1 + 2qm^+(-\frac{2\pi}{3}) \\ 1 + 2qm^+(-\frac{2\pi}{3}) & 1 + 2qm^+(-\frac{4\pi}{3}) & 1 + 2qm^+(0) \end{bmatrix}, \quad (2.14)$$

$$\mathbf{M}^-(t) = \frac{\alpha_2}{3} \begin{bmatrix} 1 + 2qm^-(0) & 1 + 2qm^-(-\frac{2\pi}{3}) & 1 + 2qm^-(-\frac{4\pi}{3}) \\ 1 + 2qm^-(-\frac{2\pi}{3}) & 1 + 2qm^-(-\frac{4\pi}{3}) & 1 + 2qm^-(0) \\ 1 + 2qm^-(-\frac{4\pi}{3}) & 1 + 2qm^-(0) & 1 + 2qm^-(-\frac{2\pi}{3}) \end{bmatrix}, \quad (2.15)$$

$$m^+ = \cos(\omega_m t + x), \quad m^- = \cos(-(\omega_m + 2\omega)t + x), \quad \omega_m = \omega_L - \omega, \quad (2.16)$$

$$\alpha_1 = \frac{1}{2}\left(1 + \frac{\tan(\varphi_S)}{\tan(\varphi_L)}\right), \quad \alpha_2 = \frac{1}{2}\left(1 - \frac{\tan(\varphi_S)}{\tan(\varphi_L)}\right). \quad (2.17)$$

Considering only the solution (2.14) ($\alpha_1 = 1$, $\alpha_2 = 0$), the phase displacement at the input is the same as at the output because $\varphi_S = \varphi_L$, whereas the solution (2.15) ($\alpha_1 = 0$, $\alpha_2 = 1$), yields $\varphi_S = -\varphi_L$ giving reversed phase displacement at the input. If both solutions are combined (2.13), the result provides the means for input displacement factor control [142]. If $\alpha_1 = \alpha_2$ the input displacement factor at the converter terminals is unity, regardless of the loads character (load displacement factor). Through the choice of α_1 and α_2, there are the possibility to input displacement factor control [110, 142].

The solution presented by Eqs. (2.13)–(2.17) is characterised as a limitation of voltage transfer ratio q. In this approach, during each switch sequence time (T_{Seq}), the average load voltage is equal to the target voltage. For this to be possible it is clear that the target voltages must fit within the source voltage envelope for any load frequency (Fig. 2.43). Then, the voltage ratio is limited to $q_{max} = 0.5$ [3, 4].

An improvement in the achievable voltage ratio to 0.866 ($\sqrt{3}/2$) is possible by adding common mode voltages to the target load voltages, as defined by Eq. (2.18) and as shown in Fig. 2.44. The matrix u_L of the target output voltages includes third harmonics of the source and load voltages. This new strategy is known as Venturini's optimum or improved method. The general conception of the improved Venturini control strategy is presented in research papers [3] and [4].

$$
\mathbf{u}_L = \begin{bmatrix} U_{L1}\cos(\omega_L t) + U_{S1}\frac{\cos(3\omega t)}{4} - U_{L1}\frac{\cos(3\omega_L t)}{6} \\ U_{L2}\cos(\omega_L t + \frac{2\pi}{3}) + U_{S2}\frac{\cos(3\omega t)}{4} - U_{L2}\frac{\cos(3\omega_L t)}{6} \\ U_{L3}\cos(\omega_L t + \frac{4\pi}{3}) + U_{S3}\frac{\cos(3\omega t)}{4} - U_{L3}\frac{\cos(3\omega_L t)}{6} \end{bmatrix}. \tag{2.18}
$$

According to [3] and [4] the transfer matrix in Venturini's improved modulation is described as follows:

$$
\mathbf{M}(t) = \begin{bmatrix} m(0,\,0,\,0,\,0,\,0,\,0)\ m(2,\,4,\,2,\,4,\,2,\,4)\ m(4,\,2,\,4,\,2,\,4,\,2) \\ m(2,\,2,\,0,\,0,\,0,\,0)\ m(4,\,0,\,2,\,4,\,2,\,4)\ m(0,\,4,\,4,\,2,\,4,\,2) \\ m(4,\,4,\,0,\,0,\,0,\,0)\ m(0,\,2,\,2,\,4,\,2,\,4)\ m(2,\,0,\,4,\,2,\,4,\,2) \end{bmatrix},
$$
$$\tag{2.19}$$

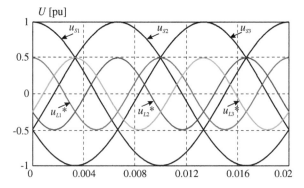

Fig. 2.43 Illustrating maximum voltage ratio of 0.5 for classical Venturini modulation

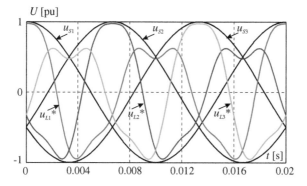

Fig. 2.44 Illustrating maximum voltage ratio of 0.866 for Venturini's improved modulation for $f_L = 100\,\text{Hz}$

Fig. 2.45 Voltage transfer ratio in MC with Venturini's optimum modulation as a function of input displacement factor

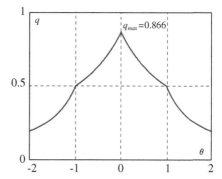

$$m(x_1, x_2, x_3, x_4, x_5, x_6)$$
$$= \frac{1}{3}\left\{1 + \frac{\sqrt{3}}{2}p\left[Z_1^1(x_1) + Z_1^{-1}(x_2) - \frac{1}{6}Z_3^1(x_3) - \frac{1}{6}Z_3^{-1}(x_4)\right.\right.$$
$$\left.+ \text{sign}(p)\left(-\frac{1}{6\sqrt{3}}Z_0^4(x_5) + \frac{7}{6\sqrt{3}}Z_0^2(x_6)\right)\right] + a_1 Z_1^1(x_1) + a_2 Z_1^{-1}(x_2)\right\},$$

$$(2.20)$$

$$Z_\alpha^\beta(\gamma, t) = \cos\left((\alpha\omega_L + \beta\omega)t + \gamma\frac{\pi}{3}\right),\qquad(2.21)$$

$$a = 2|\theta|q,\qquad p = \frac{1}{\sqrt{3}}(2q - a),\qquad \theta = \frac{\tan(\varphi_S)}{\tan(\varphi_L)},\qquad(2.22)$$

and: ($a_1 = a$ i $a_2 = 0$) where $\theta < 0$, ($a_2 = a$ i $a_1 = 0$) where $\theta > 0$, ($a_1 = a_2$) where $\theta = 0$.

An input displacement factor can be a control which uses Eqs. (2.18)–(2.22). Unfortunately, if the input displacement factor is different from unity, the voltage ratio limit will be reduced from 0.866 to a small value, which depends on the displacement factor achieved in the input site. The maximum voltage ratio is described by Eq. (2.23) and is shown in Fig. 2.45.

$$2q\left[|\theta|\left(1 - \frac{\text{sign}(p)}{\sqrt{3}}\right) + \frac{\text{sign}(p)}{\sqrt{3}}\right] \leq 1, \tag{2.23}$$

where: $\text{sign}(p) = \begin{cases} 1, & p \geq 1 \\ -1, & p \leq 0 \end{cases}$.

Scalar modulation methods

A second type of MC modulation based on a low-frequency transfer matrix is the scalar method. The basic rules of this control were first proposed by Roy, in 1987 [113]. In the proposed modulation method the switch actuation signals are calculated directly from measurements of the source phase voltages. According to [111–113], the value of any instantaneous load phase voltage may be expressed by the following equations:

$$u_L = U_{Lm}\cos(\omega_L t) = \frac{1}{T_{\text{Seq}}}(t_K u_K + t_L u_L + t_M u_M), \tag{2.24}$$

$$t_K + t_L + t_M = T_{\text{Seq}}, \tag{2.25}$$

where K–L–M are names of subscripts which change according to the rules below:

Rule 1: At any instant, the source phase voltage which has a polarity different from both others is assigned to "M".
Rule 2: The two source phase voltages which share the same polarity are assigned to "K" and "L", the smallest one of the two (in absolute value), being "K".

Then t_K and t_L are chosen such that:

$$\frac{t_K}{t_L} = u_K u_L = \rho_{KL} \qquad 0 \leq \rho_{KL} \leq 1. \tag{2.26}$$

In a balanced three-phase system, the pulse duty factors are given as:

$$\begin{aligned} m_{jL} &= \frac{t_L}{T_{\text{Seq}}} = \frac{(u_j - u_M)u_L}{1.5U_{Sm}^2} \\ m_{jK} &= \frac{t_K}{T_{\text{Seq}}} = \frac{(u_j - u_M)u_K}{1.5U_{Sm}^2} \\ m_{jM} &= \frac{t_K}{T_{\text{Seq}}} = 1 - \frac{t_K + t_L}{T_{\text{Seq}}} = 1 - (m_{jL} + m_{jK}) \end{aligned}, \tag{2.27}$$

where: $j = \{a, b, c\}$ which is the name of load phase.

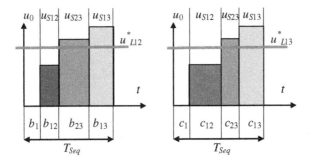

Fig. 2.46 Synthesising of output line-to-line voltages by three line-to-line source voltages in MC witch scalar modulation

The presented MC scalar modulation provides a voltage transfer ratio equal to 0.5 similar to Venturini modulation. Again, common mode addition is used with the target output voltages, to achieve a 0.866 voltage ratio capability.

Development of the Roy concept is presented by Ishiguro et al. in [65]. In the proposed modulation method the switch actuation signals are calculated directly from measurements of the line-to-line source voltages and from the demands of the load line-to-line voltages. The requirements of the output line-to-line voltages u^*_{L12} and u^*_{L13} are synthesised in each sequence period T_{Seq} by using the three input line-to-line voltages u_{S12}, u_{S23}, u_{S31} and zero voltage u_0 as follows (Fig. 2.46):

$$u^*_{L12} = b_{12}u_{S12} + b_{23}u_{S23} + b_{13}u_{S13} + b_1 u_0$$
$$u^*_{L13} = c_{12}u_{S12} + c_{23}u_{S23} + c_{13}u_{S13} + c_1 u_0 \qquad (2.28)$$

where:

$$b_{12} + b_{23} + b_{13} + b_1 = 1$$
$$c_{12} + c_{23} + c_{13} + c_1 = 1 \qquad (2.29)$$

and $0 \le b_{12} \le 1, 0 \le b_{23} \le 1, 0 \le b_{13} \le 1, 0 \le b_1 \le 1, 0 \le c_{12} \le 1, 0 \le c_{23} \le 1,$ $0 \le c_{13} \le 1, 0 \le c_1 \le 1$.

The values of coefficients are defined by (2.30)–(2.32). In this MC scalar modulation method the achieved voltage transfer ratio is equal to 0.75.

$$b_{12} = \frac{u_{S12}u^*_{L12}}{u^2_{S12} + u^2_{S23} + u^2_{S13}}, \quad b_{23} = \frac{u_{S23}u^*_{L12}}{u^2_{S12} + u^2_{S23} + u^2_{S13}},$$
$$b_{13} = \frac{u_{S13}u^*_{L12}}{u^2_{S12} + u^2_{S23} + u^2_{S13}}, \qquad (2.30)$$

$$c_{12} = \frac{u_{S12}u^*_{L13}}{u^2_{S12} + u^2_{S23} + u^2_{S13}}, \quad c_{23} = \frac{u_{S23}u^*_{L13}}{u^2_{S12} + u^2_{S23} + u^2_{S13}},$$

$$c_{13} = \frac{u_{S13}u^*_{L13}}{u^2_{S12} + u^2_{S23} + u^2_{S13}}, \tag{2.31}$$

$$b_1 = 1 - b_{12} - b_{23} - b_{13}, \quad c_1 = 1 - c_{12} - c_{23} - c_{13}. \tag{2.32}$$

When only two line-to-lines are taken into account for calculation of control signals the achieved voltage transfer ratio is equal to 0.866, and according to [65] the control strategy is described as follows:

$$\begin{aligned} u^*_{L12} &= b_2 u_{S12} + b_3 u_{S13} + b_1 u_0 \\ u^*_{L13} &= c_2 u_{S12} + c_3 u_{S13} + c_1 u_0 \end{aligned}, \tag{2.33}$$

$$b_1 + b_2 + b_3 = 1, \quad c_1 + c_2 + c_3 = 1, \tag{2.34}$$

where: $0 \le b_1 \le 1, 0 \le b_2 \le 1, 0 \le b_3 \le 1, 0 \le c_1 \le 1, 0 \le c_2 \le 1, 0 \le c_3 \le 1$. The values of coefficient are defined by Eqs. (2.35) and (2.36). Synthesis of the output line-to-line voltages in sample sequence period T_{Seq}, are shown in Fig. 2.47.

$$b_2 = \frac{(u_{S12} - u_{S23})u^*_{L12}}{u^2_{S12} + u^2_{S23} + u^2_{S13}}, \quad b_3 = \frac{(u_{S23} - u_{S31})u^*_{L12}}{u^2_{S12} + u^2_{S23} + u^2_{S13}}, \quad b_1 = 1 - b_2 - b_3,$$

$$\tag{2.35}$$

$$b_2 = \frac{(u_{S12} - u_{S23})u^*_{L13}}{u^2_{S12} + u^2_{S23} + u^2_{S13}}, \quad b_3 = \frac{(u_{S23} - u_{S31})u^*_{L13}}{u^2_{S12} + u^2_{S23} + u^2_{S13}}, \quad c_1 = 1 - c_2 - c_3.$$

$$\tag{2.36}$$

These scalar methods have limitations in input displacement factor control (input power factor). A comprehensive treatment of both voltage transfer ratio and input power factor aspects of the scalar method is contained in [106]. The maximum voltage transfer ratio is also equal to 0.866, but with a wider range of input displacement factor control.

Indirect modulation

The idea of an undirected control of a matrix converter makes the series connection of rectifier and inverter an equivalent circuit to a matrix converter, and is shown in Fig. 2.48. This technique, proposed in 1983 [109], consists of a simple control strategy where the most positive and the most negative input voltages, called here u_p and u_n, respectively, are used to synthesize the output reference voltage. In this equivalent circuit model, all kinds of significant pulse width modulation (PWM) algorithms for the rectifier and inverter can contribute to the control of the matrix converter. This concept is also known as modulation with "fictitious DC link" [142]. Then, the modulation with indirect transfer function was developed by Ziogas et al. [155, 156] and Huber et al. [61, 62].

The modulation process is divided into two steps. To attain the above features, a mathematical approach is employed as indicated in (2.37) [142].

$$\bar{\mathbf{u}}_L = (\mathbf{A}\mathbf{u}_I)\mathbf{B},\qquad(2.37)$$

where:

$$\mathbf{A} = K_A \begin{bmatrix} \cos(\omega t) \\ \cos(\omega t - 120°) \\ \cos(\omega t - 240°) \end{bmatrix}^T, \qquad \mathbf{B} = K_B \begin{bmatrix} \cos(\omega_L t) \\ \cos(\omega_L t - 120°) \\ \cos(\omega_L t - 240°) \end{bmatrix}. \qquad(2.38)$$

According to (2.38) and (2.38), after some rearranging there is:

$$\bar{\mathbf{u}}_L = \frac{3 K_A K_B U_{Sm}}{2} \begin{bmatrix} \cos(\omega_L t) \\ \cos(\omega_L t - 120°) \\ \cos(\omega_L t - 240°) \end{bmatrix}, \qquad(2.39)$$

where K_A and K_B are modulation indexes. In a simple way, the technique operation is the following. First multiplication $\mathbf{A}\mathbf{u}_S$ in (2.37) corresponds to "rectifier transformation". A "fictitious DC-link" is obtained as a result of this multiplication. Then, the second step is generally referred to as the "inverter transformation". A practical realisation of this mathematical approach is discussed in detail in [155, 156]. Generally, in the indirect modulation approach, the maximum voltage transfer ratio is equal to $q = 6\sqrt{3}/\pi^2 = 1.053$. The voltage ratio obtainable is obviously greater than that of other methods. Unfortunately, this achievement comes with low-order harmonics (low frequency distortion) of source currents, or load voltages or both. For $q < 0.866$, the indirect method yields very similar results to the direct methods.

Space vector modulation
The space vector modulation (SVM) technique is based on the instantaneous space vector representation of source and/or load voltages and/or currents in power

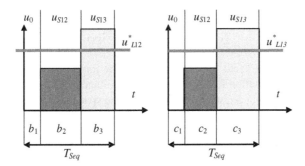

Fig. 2.47 Synthesizing of output line-to-line voltages by two line-to-line source voltages in MC with scalar modulation

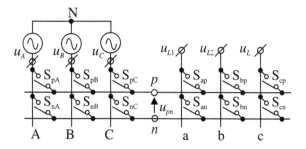

Fig. 2.48 Indirect equivalent structure of MC

converters [50, 59]. The SVM is a control technique that has been widely used in adjustable speed drives. Generally, in conventional DC-link VSI, the SVM technique is used to provide the reference output voltage vector $\overrightarrow{u}_{Lref}$. This vector is obtained by basic voltage vectors generated by the different inverter configurations. The SVM modulation for MCs is able to synthesize the reference output voltage vector $\overrightarrow{u}_{Lref}$ and due to the direct source voltages connection, it can also control the source current displacement angle [22, 121]. The SVM is probably the most used modulation strategy for MCs. Several control strategies based on the SVM technique for MCs have been proposed in the literature [17, 21–28, 54, 55, 61, 62, 121]. Basically, two methods for the implementation of SVM for MCs are used. The first one is defined as the "indirect method," because the MC is described as a two-stage system with virtual DC-link [61, 62]. The second approach to SVM for MCs is based on a direct approach [17, 21–28, 54, 55, 121]. The basic principles of direct SVM for MCs are described below.

For the space-vector modulation of the matrix converter it is convenient to define the following four space vectors [23]:

$$
\begin{aligned}
\underline{u}_O &= \tfrac{2}{3}(u_{L1} + \underline{a}u_{L2} + \underline{a}^2 u_{L3}) \\
\underline{u}_S &= \tfrac{2}{3}(u_{S1} + \underline{a}u_{S2} + \underline{a}^2 u_{S3}) \\
\underline{i}_O &= \tfrac{2}{3}(i_{L1} + \underline{a}i_{L2} + \underline{a}^2 i_{L3}) \\
\underline{i}_S &= \tfrac{2}{3}(i_{S1} + \underline{a}i_{S2} + \underline{a}^2 i_{S3})
\end{aligned}
\tag{2.40}
$$

where u_S is the space-vector representation for the input phase voltage, u_O is the space-vector representation for the output phase voltage, i_S is the space-vector representation for the input phase current and i_O is the space-vector representation for the output phase current and $\underline{a} = e^{j(2\pi/3)}$.

Taking into account that in MCs the input phases must never be short-circuited and the output currents must never be interrupted, there are 27 possible switching configurations [21]. These switching states and the resulting output voltages and source current are tabulated in Table 2.5. These combinations are depicted in Fig. 2.49 [54, 55].

Table 2.5 Switching configuration for a MC and resulting output voltages and source current

No.	Vector No.	a	b	c	S_{aA}	S_{aB}	S_{aC}	S_{bA}	S_{bB}	S_{bC}	S_{cA}	S_{cB}	S_{cC}	u_{ab}	u_{bc}	u_{ca}	i_A	i_B	i_C
1	0_A	A	A	A	1	0	0	1	0	0	1	0	0	0	0	0	0	0	0
2	0_B	B	B	B	0	1	0	0	1	0	0	1	0	0	0	0	0	0	0
3	0_C	C	C	C	0	0	1	0	0	1	0	0	1	0	0	0	0	0	0
4	-3	A	C	C	1	0	0	0	0	1	0	0	1	$-u_{CA}$	0	u_{CA}	i_a	0	$-i_a$
5	$+2$	B	C	C	0	1	0	0	0	1	0	0	1	u_{BC}	0	$-u_{BC}$	0	i_a	$-i_a$
6	-1	B	A	A	0	1	0	1	0	0	1	0	0	$-u_{AB}$	0	u_{AB}	$-i_a$	i_a	0
7	$+3$	C	A	A	0	0	1	1	0	0	1	0	0	u_{CA}	0	$-u_{CA}$	$-i_a$	0	i_a
8	-2	C	B	B	0	0	1	0	1	0	0	1	0	$-u_{BC}$	0	u_{BC}	0	$-i_a$	i_a
9	$+1$	A	B	B	1	0	0	0	1	0	0	1	0	u_{AB}	0	$-u_{AB}$	i_a	$-i_a$	0
10	-6	C	A	C	0	0	1	1	0	0	0	0	1	u_{CA}	$-u_{CA}$	0	i_b	0	$-i_b$
11	$+5$	C	B	C	0	0	1	0	1	0	0	0	1	$-u_{BC}$	u_{BC}	0	0	i_b	$-i_b$
12	-4	A	B	A	1	0	0	0	1	0	1	0	0	u_{BA}	$-u_{BA}$	0	$-i_b$	i_b	0
13	$+6$	A	C	A	1	0	0	0	0	1	1	0	0	$-u_{CA}$	u_{CA}	0	$-i_b$	0	i_b
14	-5	B	C	B	0	1	0	0	0	1	0	1	0	u_{BC}	$-u_{BC}$	0	0	$-i_b$	i_b
15	$+4$	B	A	B	0	1	0	1	0	0	0	1	0	$-u_{AB}$	u_{AB}	0	i_b	$-i_b$	0
16	-9	C	C	A	0	0	1	0	0	1	1	0	0	0	u_{CA}	$-u_{CA}$	i_c	0	$-i_c$
17	$+8$	C	C	B	0	0	1	0	0	1	0	1	0	0	$-u_{BC}$	u_{BC}	0	i_c	$-i_c$
18	-7	A	A	B	1	0	0	1	0	0	0	1	0	0	u_{AB}	$-u_{AB}$	$-i_c$	i_c	0
19	$+9$	A	A	C	1	0	0	1	0	0	0	0	1	0	$-u_{CA}$	u_{CA}	$-i_c$	0	i_c
20	-8	B	B	C	0	1	0	0	1	0	0	0	1	0	u_{BC}	$-u_{BC}$	0	$-i_c$	i_c
21	$+7$	B	B	A	0	1	0	0	1	0	1	0	0	0	$-u_{AB}$	u_{AB}	i_c	$-i_c$	0
22	$-$	A	B	C	1	0	0	0	1	0	0	0	1	u_{AB}	u_{BC}	u_{CA}	i_a	i_b	i_c
23	$-$	A	C	B	1	0	0	0	0	1	0	1	0	$-u_{CA}$	$-u_{BC}$	$-u_{AB}$	i_a	i_c	i_b
24	$-$	B	A	C	0	1	0	1	0	0	0	0	1	$-u_{AB}$	$-u_{CA}$	$-u_{BC}$	i_b	i_a	i_c
25	$-$	B	C	A	0	1	0	0	0	1	1	0	0	u_{BC}	u_{CA}	u_{AB}	i_c	i_a	i_b
26	$-$	C	A	B	0	0	1	1	0	0	0	1	0	u_{CA}	u_{AB}	u_{BC}	i_b	i_c	i_a
27	$-$	C	B	A	0	0	1	0	1	0	1	0	0	$-u_{BC}$	$-u_{AB}$	$-u_{CA}$	i_c	i_b	i_a

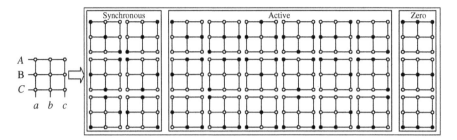

Fig. 2.49 Allowed switch combinations in a MC

Analyzing the Table 2.5, the switch configuration may be categorised as one of
the three groups [21].

Group 1 – 3 combinations giving null output voltage and input current vectors,
will be named "zero configurations". All three output phases are connected to the
same input phase in these combinations.

Group 2 – 6 combinations in which each output phase is connected to a different
input phase, will be named "synchronous configurations". In this case, the output
voltage and input current vectors have variable directions and cannot be usefully
used to synthesise the reference vectors.

Group 3 – 18 combinations where the output voltage and the input current vectors
have fixed directions and will be named "active configurations." The magnitude
of these vectors depends upon the instantaneous values of the input line-to-line
voltages and output line currents, respectively. In this case, two output lines are
connected to the same input line.

In Fig. 2.50 the output voltage and input current vectors corresponding to the 18
active configurations are shown. In this figure, the scheme of how the complex space
vector plane is divided into sectors is also presented. S_O denotes the sector containing
the output voltage vector and S_i denotes the sector containing the input current vector.
The active configurations are split into three sub-groups as shown in Table 2.5 and
Fig. 2.50. The configurations in each sub-group produce a space voltage and current
vectors in a defined direction, which change every $120°$. The amplitude and polarity
of the space vectors along the defined direction depend on which of the line-to-line
voltages is used.

The SVM algorithm for an MC is able to synthesize the reference output voltage
vector and to control the phase angle of the input current vector selecting four non-
zero configurations, which are applied for a suitable time period within each sequence
T_{Seq} as is determined by the following equation [27]:

$$\underline{u}_O = d_I \underline{u}_I + d_{II} \underline{u}_{II} + d_{III} \underline{u}_{III} + d_{IV} \underline{u}_{IV}, \tag{2.41}$$

where $\underline{u}_I$, $\underline{u}_{II}$, $\underline{u}_{III}$ and $\underline{u}_{IV}$ are the output voltage vectors corresponding to the four
selected configurations, and d_I, d_{II}, d_{III} and d_{IV} are their duty cycles, defined as:

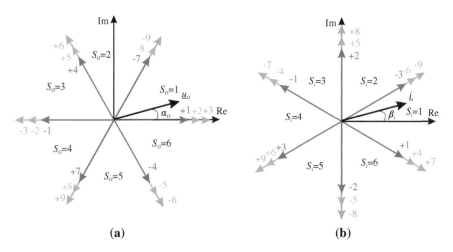

Fig. 2.50 Graphical interpretation of: **a** sectors and direction of the output voltage vectors, **b** sectors and directions of the input line current vectors

$$d_k = \frac{t_k}{T_{Seq}}, \qquad k = I, II, III, IV. \tag{2.42}$$

Finally, the zero configurations are applied to complete T_{Seq}, where:

$$d_0 = 1 - d_I - d_{II} - d_{III} - d_{IV}. \tag{2.43}$$

The rotating vectors of Group 3 are not used in SVM.

The main task of the SVM technique is to calculate the duty cycles and define the switching pattern. Several control strategies based on the SVM technique for the MC with a different sequence of the switches have been proposed in the literature [22–24, 27, 28, 54, 55, 102]. The presented types of switching sequences have been mainly focused on the possibility of reducing the number of commutations in each switching period T_{Seq} [102] and the power losses [54, 55], to improve the waveform of the output voltage and to eliminate the current distortion [22–24, 27] or to extend the operating region [28].

In order to explain the basic SVM algorithm, an example of synthesis of the reference output voltage vector and input current vector, with both lying in Sector 1, is shown in Fig. 2.51. Figure 2.51 clearly shows that the reference output voltage $\underline{u}_O$ is resolved into the components $\underline{u}'_O$ and $\underline{u}''_O$ along the two adjacent vector directions. The vector of source currents i_S is also resolved into components, along the two adjacent current directions [23]. The $\underline{u}'_O$ component can be synthesised using two voltage vectors having the same direction of u_O, whereas the $\underline{u}''_O$ component can be synthesised using two voltage vectors having the opposite direction, as follows [28]:

$$\underline{u}'_O = d_I \underline{u}_I + d_{II} \underline{u}_{II} = 2/\sqrt{3}U_O \cos(\alpha_O - \pi/3)e^{j[(S_O-1)\pi/3+\pi/3]}$$
$$\underline{u}''_O = d_{III} \underline{u}_{III} + d_{IV} \underline{u}_{IV} = 2/\sqrt{3}U_O \cos(\alpha_O + \pi/3)e^{j[(S_O-1)\pi/3]} \,, \qquad (2.44)$$

where d_I, d_{II}, d_{III} and d_{IV} are the on-time ratios of individual switch combinations corresponding to vectors $\underline{u}_I$, $\underline{u}_{II}$, $\underline{u}_{III}$ and $\underline{u}_{IV}$, α_O is the angle of the output voltage vector measured from the bisecting line of the corresponding sectors and is defined by (2.51) [23]. The requirements of the reference input current displacement angle are defined [28]

$$(d_I \underline{i}_I + d_{II} \underline{i}_{II})je^{j\beta_i}e^{j(S_i-1)\pi/3} = 0$$
$$(d_{III} \underline{i}_{III} + d_{IV} \underline{i}_{IV})je^{j\beta_i}e^{j(S_i-1)\pi/3} = 0 \,, \qquad (2.45)$$

where β_i is the angle input current vector measured from the bisecting line of the corresponding sectors and is defined by (2.51) [23], where i_I, i_{II}, i_{III}, i_{IV} are the source current vectors corresponding to the four selected configurations.

For example from Fig. 2.51 ($S_i = 1$, $S_O = 1$), possible switching states that can be utilised to synthesise the resolved voltage and current components are

$$\underline{u}^1 : \pm 7, \pm 8, \pm 9, \quad \underline{u}^2 : \pm 1, \pm 2, \pm 3,$$
$$\underline{i}^1 : \pm 1, \pm 4, \pm 7, \quad \underline{i}^2 : \pm 3, \pm 6, \pm 9. \qquad (2.46)$$

Then, to simultaneously synthesise the output voltage and the input current vectors, common switching states ± 9, ± 7, ± 1 and ± 3 are selected. From two switching states with the same number but opposite polarity, only one should be used. If the duty cycle is positive, the switching state with a positive polarity is selected; otherwise, the one with a negative polarity is selected. The required modulation duty cycles for the switching configurations I, II, III and IV are given by Eqs. (2.47)–(2.50) [23]:

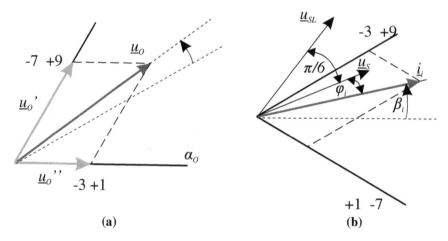

(a)   (b)

Fig. 2.51 Synthesis of: **a** reference output voltage vector, **b** reference input current vector

$$d_I = (-1)^{(S_O+S_i+1)} \frac{2}{\sqrt{3}} q \frac{\cos(\alpha_O - \frac{\pi}{3})\cos(\beta_i - \frac{\pi}{3})}{\cos(\varphi_i)}, \tag{2.47}$$

$$d_I = (-1)^{(S_O+S_i)} \frac{2}{\sqrt{3}} q \frac{\cos(\alpha_O - \frac{\pi}{3})\cos(\beta_i + \frac{\pi}{3})}{\cos(\varphi_i)}, \tag{2.48}$$

$$d_I = (-1)^{(S_O+S_i)} \frac{2}{\sqrt{3}} q \frac{\cos(\alpha_O + \frac{\pi}{3})\cos(\beta_i - \frac{\pi}{3})}{\cos(\varphi_i)}, \tag{2.49}$$

$$d_I = (-1)^{(S_O+S_i+1)} \frac{2}{\sqrt{3}} q \frac{\cos(\alpha_O + \frac{\pi}{3})\cos(\beta_i + \frac{\pi}{3})}{\cos(\varphi_i)}. \tag{2.50}$$

In (2.44)–(2.50) φ_i is the input phase displacement angle, α_O and β_i are the angles of the output voltage and input current vectors measured from the bisecting line of the corresponding sectors, and are limited as follows [23]:

$$-\frac{\pi}{6} \le \alpha_O \le \frac{\pi}{6}, \qquad -\frac{\pi}{6} \le \beta_i \le \frac{\pi}{6}. \tag{2.51}$$

Equations (2.47)–(2.50) have a general validity for any combination of the output voltage sector S_O and the input current sector S_i. Table 2.6 provides the four switch configurations to be used within the cycle period T_{Seq} [23]. The sectors are defined as shown in Fig. 2.50.

Subject to the constraints:

$$|d_I| + |d_{II}| + |d_{III}| + |d_{IV}| \le 1, \tag{2.52}$$

the voltage ratio is defined by

$$q \le \frac{\sqrt{3}|\cos(\varphi_i)|}{2\cos(\beta_i)\cos(\alpha_O)}. \tag{2.53}$$

In the particular case of balanced supply voltages and balanced output voltages, the maximum voltage transfer ratio is equal

$$q \le \frac{\sqrt{3}}{2}|\cos(\varphi_i)|. \tag{2.54}$$

This means that the maximum voltage transfer ratio of MC is equal to 0.866 if the unity input power factor is set [142].

Current source matrix converter
As mentioned previously, the MC depends on the power supply (with voltage or current character) which is used—a voltage source matrix converter (VSMC) or current source matrix converter (CSMC), respectively. In the ideal case, the current source matrix converter consists of current at the input side and voltage source at the output side. In the practical realisation of CSMC, the converter includes nine

Table 2.6 Selection of the switching configurations for each combination of output voltage and input current sectors

Sector S_i of the input current vector	Sector of the output voltage vector S_O																							
	1				2				3				4				5				6			
1	+9	−7	−3	+1	−6	+4	+9	−7	+3	−1	−6	+4	−9	+7	+3	−1	+6	−4	−9	+7	−3	+1	+6	−4
2	−8	+9	+2	−3	+5	−6	−8	+9	−2	+3	+5	−6	+8	−9	−2	+3	−5	+6	+8	−9	+2	−3	−5	+6
3	+7	−8	−1	+2	−4	+5	+7	−8	+1	−2	−4	+5	−7	+8	+1	−2	+4	−5	−7	+8	−1	+2	+4	−5
4	−9	+7	+3	−1	+6	−4	−9	+7	−3	+1	+6	−4	+9	−7	−3	+1	−6	+4	+9	−7	+3	−1	−6	+4
5	+8	−9	−2	+3	−5	+6	+8	−9	+2	−3	−5	+6	−8	+9	+2	−3	+5	−6	−8	+9	−2	+3	+5	−6
6	−7	+8	+1	−2	+4	−5	−7	+8	−1	+2	+4	−5	+7	−8	−1	+2	−4	+5	+7	−8	+1	−2	−4	+5
Index of d_k	I	II	III	IV	I	II	III	IV	I	II	III	IV	I	II	III	IV	I	II	III	IV	I	II	III	IV

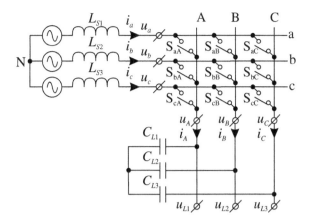

Fig. 2.52 Current source matrix converter

bidirectional switch cells and three AC capacitors that are located at the load-side of the converter. The current source realization includes a three-phase ideal voltage source (u_{S1}, u_{S2}, u_{S3}) in series with an R–L impedance per phase. Figure 2.52 shows the topology of a CSMC. In the literature, the CSMC is not often considered. Only a few papers present the principle of CSMC [48, 49, 70, 87, 88, 103].

The relationship between the converter source-side and load-side currents and voltages is

$$\begin{bmatrix} i_A(t) \\ i_B(t) \\ i_C(t) \end{bmatrix} = \begin{bmatrix} s_{aA}(t) & s_{bA}(t) & s_{cA}(t) \\ s_{aB}(t) & s_{bB}(t) & s_{cB}(t) \\ s_{aC}(t) & s_{bC}(t) & s_{cC}(t) \end{bmatrix} \begin{bmatrix} i_a(t) \\ i_b(t) \\ i_c(t) \end{bmatrix}, \tag{2.55}$$

$$\begin{bmatrix} u_a(t) \\ u_b(t) \\ u_c(t) \end{bmatrix} = \begin{bmatrix} s_{aA}(t) & s_{aB}(t) & s_{aC}(t) \\ s_{bA}(t) & s_{bB}(t) & s_{bC}(t) \\ s_{cA}(t) & s_{cB}(t) & s_{cC}(t) \end{bmatrix} \begin{bmatrix} u_A(t) \\ u_B(t) \\ u_C(t) \end{bmatrix}. \tag{2.56}$$

The main objective of CSMC is to control directly the magnitude, frequency and phase-angle of the load current. Furthermore, indirectly, the output voltage is controlled. It is possible to obtain a voltage gain greater than one. The voltage gain strictly depends on the load. The CSMC also controls the phase-angle of the voltages $(u_a, u_b$ and $u_c)$ at the input side of the matrix switches. In this way, it is possible to control the input power factor.

The major disadvantage of the CSMC is the realisation of a practical current source—a large inductance value is needed. Moreover, the energy accumulated in source inductors is dangerous in the case when turning-off all converter switches. An additional circuit to discharge this energy is required. Another drawback of CSMC is that the voltage gain is dependent on load change. In large loads, the output voltages can be much smaller than the source one.

On the load side, the CSMC can be considered as a voltage source MC. Due to this fact, in the CSMC there can be used, with small modifications, all of the modulation methods, commutation strategies and protection strategies which are used in the VSMC. An exemplary solution of a CSMC with a Venturini control strategy is presented in [87], in which the low-frequency modulation matrix is transposed in accordance with classical Venturini modulation [135]. In the solution of a CSMC with SVM [48], the voltage sectors are dependent on input voltages (u_a, u_b and u_c), and current sectors are calculated from output current, inversely, as in the VSMC. The commutation method is based on input current measurements [48].

The matrix of nine bi-directional switches with capacitors connected on the input side terminals is versatile. Depending on the source character—voltage or current—the source can be connected to the input or output sides of the matrix switches. In this way, we obtain the universal converter which is connected between voltage and current sources, and which can transfer energy in both directions. These beneficial properties of matrix-connected switches are introduced in a new family of matrix-reactance frequency converters, which are the main goal of this book and are presented in the following chapters.

Multilevel matrix converter

As mentioned in the introduction, the multilevel concept of a direct matrix converter is also proposed. It is well-known that multilevel technology is a good solution in medium or high voltage power conversion [59]. The multilevel matrix converter (MLMC) is obtained by replacing each switch in a direct MC by two or more series connected switches, and flying capacitors which are introduced to clamp the voltage over the switches as shown in Fig. 2.53 [120, 143, 147]. Similarly to a MC, the MLMC must fulfil the same constraints on the switching pattern to avoid short circuit on the input side or open circuits on the output side. An additional problem is voltage control across the flying capacitor which changes periodically in line with the input voltage. The voltages must be controlled to keep a sinusoidal shape. Because of this, the modulation methods and commutation strategies are complex. This kind of converter is not a classical direct AC–AC frequency converter without DC energy storage elements, because the flying capacitors are used as a local energy storage. The advantages of MLMC is the possibility to apply it in high voltage range systems with low maintenance costs and few voltage device components. Furthermore, the MLMC gives an improved output voltage waveform quality. The maximum voltage transfer ratio in a MLMC is less than one. In the case with Venturini modulation, it is equal to 0.5 [120, 147] and with SVM it is equal to 0.8 [91, 114, 146].

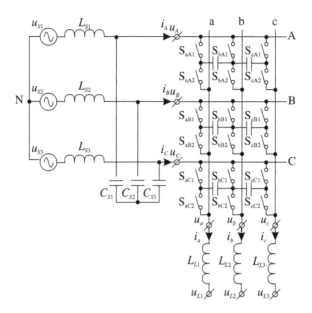

Fig. 2.53 Three level matrix converter

2.3.3 Indirect AC–AC Frequency Converters Without DC Energy Storage Elements

The second group of frequency converters without DC energy storage elements is the indirect converter with fictitious DC link (Fig. 2.2). This converter is obtained from the classical matrix converter structure. Systems with MC modulation schemes can be classified under direct frequency conversion schemes [21, 126, 135, 142] and indirect frequency conversion schemes [61, 62, 155, 156]. In the latter modulation concept, the converter is divided into two parts with fictitious DC link. There is a fictitious voltage-fed rectifier on the input side and a fictitious voltage source inverter on the output side. The input rectifier and output inverter are directly connected on the DC side. The indirect matrix converters with fictions DC link is a hardware implementation of this basic idea. Such a converter was suggested in [64] and investigated in more detail in [63, 68, 69, 76]. Figure 2.54 shows the circuit of indirect matrix converters (IMC). The main circuit consists of the PWM rectifier section (CSR—current source rectifier), PWM-inverter section (VSI) and the AC source filter. The cascaded connection of two bridge converters (CSR and VSI) provides separation of the modulation process on the input and output sides. The synchronisation of the modulation process of CSR and VSI is necessary for system balance with sinusoidal supply current because of the absence of the DC energy storage element. The IMC offers the same benefits and disadvantages as the classical direct MC, and is also called a two-stage matrix converter.

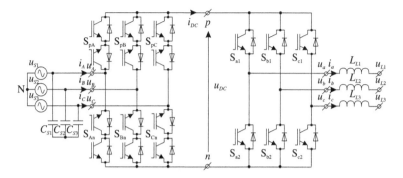

Fig. 2.54 Indirect matrix converter

MCs are inherently bi-directional and therefore can regenerate energy back into the mains from the load side. However, the DC voltage in an IMC has only positive polarity. In order to allow bi-directional current flow (four-quadrant operation), in the input bridge of the IMC bi-directional active switches are needed as shown in Fig. 2.54. On the output side, the classical voltage source inverter (Fig. 2.3) [59, 72] is used to form output voltages. The IMC employs 18 IGBTs and 18 diodes similar to those in the classical direct MC. However, the physical realisation is much easier, because the inverter stage could be realised by a conventional six-pack power module as compared to a fully discrete realisation of a direct MC.

Similarly as in the conventional MC, commutation methods are needed to avoid shorting of input phases without cutting load current path. The four-step commutation is commonly used in the IMC. The modulation strategies are based on an indirect concept which is presented in [61, 62, 155, 156]. Both vector and triangular wave modulation have been developed and are presented in [63, 64, 68, 69, 76]. In the modulation strategies, the 72 basic switch configurations are used. In Table 2.7, output voltages and source currents resulting from the different switch combinations are tabulated [81].

In Ref. [81] the authors show new topologies of IMC with reduced number of active switches on the line bridge. The input stage of the IMC is realised with four-quadrant switches, as shown in Fig. 2.54. In this circuit configuration, the bi-directional power flow could in principle be obtained with positive and negative DC link voltage. If assumed that the DC link voltage is positive polarity, a reduction in the number of active switch devices is possible. The derivation of a simplified bridge branch structure of the IMC is shown in Fig. 2.55. The new topology is called by the authors a sparse matrix converter (SMC), and is shown in Fig. 2.56. A detailed description of derivation is presented in [81, 83, 84, 116]. SMC are functionally equivalent to IMC, but are characterised by a lower realisation effort and a lower control complexity. The SMC topology employs 15 IGBTs and 18 diodes.

If a unidirectional power flow is required, a more simplified version of the system is possible. Such a circuit is shown in Fig. 2.57 [81, 83, 84, 117]. This converter is

Table 2.7 Switching configuration for an IMC and resulting output voltages and source current

No.	a	b	c	S_{pA}	S_{pB}	S_{pC}	S_{An}	S_{Bn}	S_{Cn}	S_a	S_b	S_c	u_{ab}	u_{bc}	u_{ca}	u_{DC}	i_A	$i_B\;i_C$
1	p	p	p	X	X	X	X	X	1	1	1	1	0	0	–	0	0	0
10	p	n	n	X	X	X	X	X	0	0	0	0	0	0	–	0	0	0
19	X	X	X	1	0	1	0	0	1	X	X	X	0	0	0	0	0	0
25	X	X	X	0	0	0	1	0	0	X	X	X	0	0	0	0	0	0
31	X	X	X	0	1	0	0	1	0	X	X	X	0	0	0	0	0	0
37	A	C	C	0	0	1	1	0	0	0	1	1	0	u_{CA}	$-u_{CA}$	i_a	0	$-i_a$
38	A	C	C	0	0	1	1	0	0	1	0	0	0	u_{CA}	u_{CA}	i_a	0	$-i_a$
39	B	C	C	0	1	0	0	0	1	0	1	1	0	$-u_{BC}$	u_{BC}	0	i_a	$-i_a$
40	B	C	C	0	1	0	0	0	1	1	0	0	0	$-u_{BC}$	$-u_{BC}$	0	i_a	$-i_a$
41	B	A	A	1	0	0	0	1	0	0	1	1	0	u_{AB}	$-u_{AB}$	$-i_a$	i_a	0
42	B	A	A	1	0	0	0	1	0	1	0	0	0	u_{AB}	u_{AB}	$-i_a$	i_a	0
43	C	A	A	0	0	1	1	0	0	0	1	1	0	$-u_{CA}$	u_{CA}	$-i_a$	0	i_a
44	C	A	A	0	0	1	1	0	0	1	0	0	0	$-u_{CA}$	$-u_{CA}$	$-i_a$	0	i_a
45	C	B	B	0	1	0	0	0	1	0	1	1	0	u_{BC}	$-u_{BC}$	0	$-i_a$	i_a
46	C	B	B	0	1	0	0	0	1	1	0	0	0	u_{BC}	u_{BC}	0	$-i_a$	i_a
47	A	B	B	1	0	0	0	1	0	0	1	1	0	$-u_{AB}$	u_{AB}	i_a	$-i_a$	0
48	A	B	B	1	0	0	0	1	0	1	0	0	0	$-u_{AB}$	$-u_{AB}$	i_a	$-i_a$	0
49	C	A	C	0	0	1	1	0	0	0	1	1	$-u_{CA}$	0	$-u_{CA}$	i_b	0	$-i_b$
50	C	A	C	0	0	1	1	0	0	1	0	0	$-u_{CA}$	0	u_{CA}	i_b	0	$-i_b$
51	C	B	C	0	1	0	0	0	1	1	0	0	u_{BC}	0	u_{BC}	0	i_b	$-i_b$
52	C	B	C	0	1	0	0	0	1	0	1	1	u_{BC}	0	$-u_{BC}$	0	i_b	$-i_b$
53	A	B	A	1	0	0	0	1	0	0	1	1	$-u_{AB}$	0	$-u_{AB}$	$-i_b$	i_b	0
54	A	B	A	1	0	0	0	1	0	1	0	0	$-u_{AB}$	0	u_{AB}	$-i_b$	i_b	0
55	A	C	A	0	0	1	1	0	0	1	0	0	u_{CA}	0	u_{CA}	$-i_b$	0	i_b
56	A	C	A	0	0	1	1	0	0	0	1	1	u_{CA}	0	$-u_{CA}$	$-i_b$	0	i_b
57	B	C	B	0	1	0	0	0	1	1	0	0	$-u_{BC}$	0	$-u_{BC}$	$-i_b$	i_b	0

(continued)

Table 2.7 (continued)

No.	a	b	c	S_{pA}	S_{pB}	S_{pC}	S_{An}	S_{Bn}	S_{Cn}	S_a	S_b	S_c	u_{ab}	u_{bc}	u_{ca}	u_{DC}	i_A	i_B	i_C
58	B	C	B	0	1	0	1	0	1	0	1	u_{BC}	$-u_{BC}$	0	u_{BC}	0	$-i_b$	i_b	i_b
59	B	A	B	0	0	0	1	0	0	1	0	$-u_{AB}$	u_{AB}	0	u_{AB}	i_b	$-i_b$	0	0
60	B	A	B	1	0	1	0	1	1	0	1	$-u_{AB}$	u_{AB}	0	$-u_{AB}$	i_b	$-i_b$	0	0
61	C	C	A	0	0	0	0	1	0	0	1	0	u_{CA}	$-u_{CA}$	$-u_{CA}$	i_c	0	$-i_c$	$-i_c$
62	C	C	A	0	1	1	0	1	1	1	0	0	u_{CA}	$-u_{CA}$	u_{CA}	i_c	0	$-i_c$	$-i_c$
63	C	C	B	0	0	0	1	0	0	0	1	0	$-u_{BC}$	u_{BC}	u_{BC}	0	i_c	$-i_c$	$-i_c$
64	C	C	B	1	0	1	0	1	1	1	0	0	$-u_{BC}$	u_{BC}	$-u_{BC}$	0	i_c	$-i_c$	$-i_c$
65	A	A	B	0	0	0	0	1	0	0	1	0	u_{AB}	$-u_{AB}$	$-u_{AB}$	$-i_c$	i_c	0	0
66	A	A	B	0	1	1	0	1	1	1	0	0	u_{AB}	$-u_{AB}$	u_{AB}	$-i_c$	i_c	0	0
67	A	A	C	0	0	0	1	0	0	0	1	0	$-u_{CA}$	u_{CA}	u_{CA}	$-i_c$	i_c	i_c	i_c
68	A	A	C	1	0	1	0	1	1	1	0	0	$-u_{CA}$	u_{CA}	$-u_{CA}$	$-i_c$	0	i_c	i_c
69	B	B	C	0	0	0	0	1	0	0	1	0	u_{BC}	$-u_{BC}$	$-u_{BC}$	0	$-i_c$	i_c	i_c
70	B	B	C	0	1	1	0	1	1	1	0	0	u_{BC}	$-u_{BC}$	u_{BC}	0	$-i_c$	i_c	i_c
71	B	B	A	0	0	0	1	0	0	0	1	0	$-u_{AB}$	u_{AB}	u_{AB}	i_c	$-i_c$	0	0
72	B	B	A	1	0	1	0	1	1	1	0	0	$-u_{AB}$	u_{AB}	$-u_{AB}$	i_c	$-i_c$	0	0

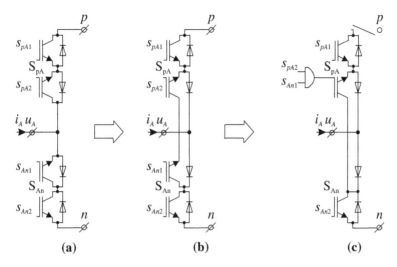

Fig. 2.55 Derivation of the bridge branch **a** branch for IMC, **b** idea of reduction of branch transistors, **c** branch for SMC

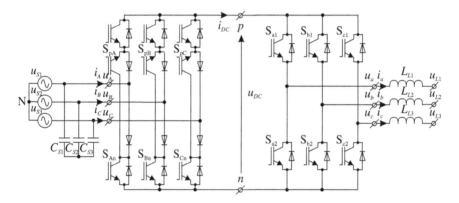

Fig. 2.56 Sparse matrix converter (SMC)

named as an ultra sparse matrix converter (USMC). The source bridge consists of only three IGBTS and 12 diodes.

The structures of IMC, SMC and USMC require the use of the multistep switch commutation method to fulfil commutation rules. In another topology previously presented in paper [138], a simplified commutation is possible. This converter is named a very sparse matrix converter (VSMC) and is shown in Fig. 2.58. The rectifier bridge consists of bi-directional switches with IGBT and diode bridge configuration. This switching connection provides the zero DC-link current commutation strategy [81]. In VSMC, only a safety interval dead time is required between switch-off of one four-quadrant switch and the switch-on of the next four-quadrant switch. This topology provides an option to reduce the number of IGBTs, but the number of power

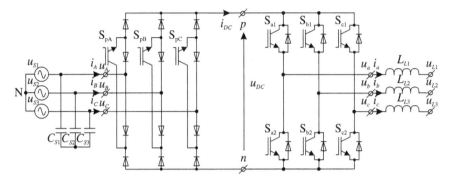

Fig. 2.57 Ultra sparse matrix converter (USMC)

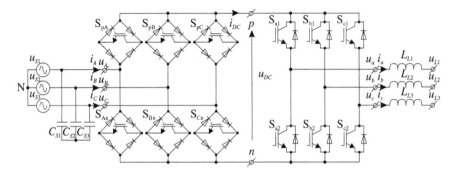

Fig. 2.58 Very sparse matrix converter (VSMC)

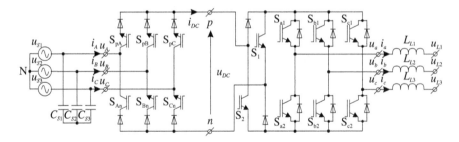

Fig. 2.59 Inversing link matrix converter (ILMC)

diodes is increased. In the VSMC structure, the 12 IGBTs and 30 diodes is employed. This topology is functionally similar to IMC.

Zero DC link current commutation and bidirectional power flow also would allow the employment of the circuit topology of the inverting link matrix converter (ILMC) presented in [81] and shown in Fig. 2.59. Here, the bidirectional current carrying capability of the input stage is achieved by connecting an input rectifier and a voltage inverter through two power transistors and two diodes. Unfortunately, the inversion of the voltages has to be performed with high frequency. Then the switching losses are high. Additionally the modulation process is complex. Therefore, the literature on the ILMC will not be considered in more detail.

The topologies of the sparse, very sparse, ultra sparse and inverting link matrix converter are characterised by a voltage transfer ratio also less than one, with its maximal level equal to 0.866.

Indirect matrix converters have also been extended into multilevel structures. Multilevel indirect matrix converters (MIMC) is an emerging topology that integrates the multilevel concept into the indirect matrix converter topology. Having the ability to generate multilevel output voltages, the MIMC is able to produce better quality output waveforms than IMC in terms of harmonic content, but at the cost of a higher number of power semiconductor devices requirement and more complicated modulation strategy. In the literature there are presented two main topologies, a three-level-output-stage indirect matrix converter [81, 90, 92] and an indirect three-level sparse matrix converter [89]. The first one applies the three-level neutral-point-clamped voltage source inverter concept [59] to the inversion stage of an indirect matrix converter topology and is shown in Fig. 2.60. The rectified DC-link voltage, u_{DC} is transformed into dual voltage supplies, $+u_{DC}$ and $-u_{DC}$, by connecting the DC link middle point of the three-level neutral-point-clamped voltage source inverter to the neutral-point "N" of the star-connected input filter capacitors. Then, there are three voltage levels at the DC links: $+u_{DC}$, 0 V and $-u_{DC}$. Based on these voltage levels, the inversion stage can be modulated to generate the multilevel output voltage waveforms.

The second MIMC applies simplified three-level neutral-point-clamped voltage source inverter concept with neutral-point chopper, to the inversion stage of an IMC topology (Fig. 2.61) [89]. This converter topology has a simpler circuit configuration than converter from Fig. 2.60 and is also able to generate three-level output voltages.

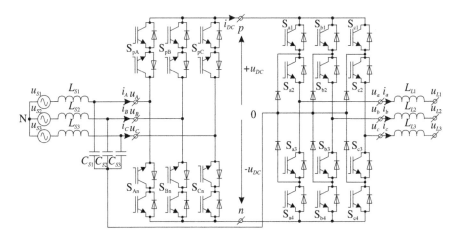

Fig. 2.60 Three-level-output-stage indirect matrix converter

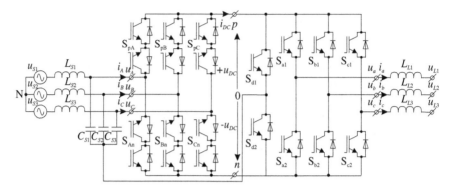

Fig. 2.61 Indirect three-level sparse matrix converter

In this circuit, two additional unidirectional switches are connected as an additional inverter leg. The main task of this leg is a device of neutral-point commutator.

2.3.4 AC–AC Frequency Converters Based on Matrix-Reactance Chopper Topologies

The first study of topologies of frequency converters based on matrix-reactance chopper topologies was presented in 1993 by Antic et al. in [8–10]. The proposed topology consists of a three phase AC–AC buck-boost chopper with an integrated matrix-converter—Fig. 2.62a. This topology has only a small regenerative AC energy storage, and was proposed for low frequency operation in drive systems with induction motors.

Another concept of frequency converters based on a MRC, but with a different switch realisation is proposed by Zinoviev et al. [104, 153, 154] (Fig. 2.62b). The principle of operation of this new voltage controller is similar to the operation of a buck-boost AC–AC chopper with the possibility to change the output voltage frequency.

Further, detailed analysis of this kind of topologies was carried out by Fedyczak et al. in a series of papers: [36–47, 85, 86, 125–129]. As a result, the new family of AC–AC frequency converters based on matrix-reactance chopper topologies was presented [45, 46, 126]. Those converters are named matrix-reactance frequency converters (MRFC) and are the main object of this book. The analysis of properties of MRFC will be presented in the following chapters. In Chap. 3, the topology generation is presented. Chapter 4 shows the modeling concept. Whereas in further chapters, there are presented the test results of theoretical analysis, simulation and experimental investigations. In this book, all the structures of frequency converters based on MRC topologies are subsequently named as matrix-reactance frequency converters.

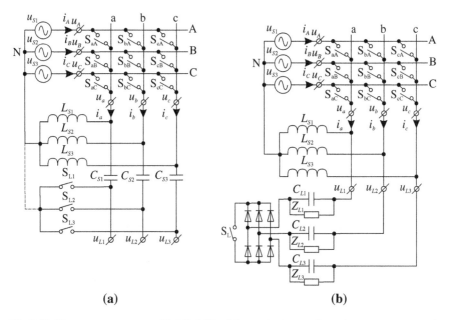

(a) **(b)**

Fig. 2.62 Frequency converters with AC–AC buck-boost choppers and matrix converter proposed: **a** in [8–10], **b** in [104, 153, 154]

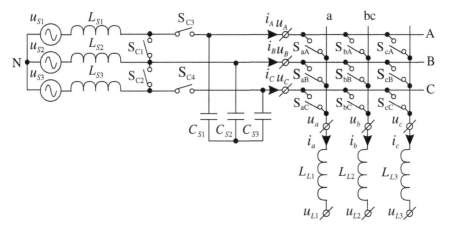

Fig. 2.63 Frequency converters with cascade connected AC–AC boost choppers and matrix converter

In recent times, another topology based on MRC has been presented, consisting of cascade connected boost MRC and MC as shown in Fig. 2.63 [66, 79, 80]. The operation of a boost chopper is only necessary when a voltage gain factor greater than 0.866 is required. If the required voltage gain is less than 0.866, then the topology is similar to the MC. The switches S_{C1} and S_{C2} are turned-off and the switches

S_{C3} and S_{C4} are turned-on. The increase of conduction losses in the case when the voltage gain is less than 0.866, is a drawback of this solution. Then, two bidirectional switches are turned-on all the time.

All frequency converters based on AC–AC matrix-reactance choppers offer a voltage transfer ratio greater than one.

2.4 Hybrid AC–AC Frequency Converters

The last group of frequency converters (Fig. 2.2) includes the hybrid solution. In order to obtain a voltage gain greater than one, the combination of AC–AC frequency converters (without DC energy storage) and one small or several small local DC energy storage elements or DC–DC buck-boost (boost) choppers are proposed. All hybrid frequency converters include a DC energy storage element, but with small dimensions. The topologies are more complex and modulation and commutation strategies are also elaborate.

In paper [35] the authors have proposed the first hybrid topology. Figure 2.64 shows a proposed circuit, which was named a modular matrix converter (MMC) [7]. The MMC is obtained by replacing each switch in the classical direct MC [135] by single-phase H bridge inverters. Unfortunately, the number of active devices increases, but in this topology has not a main DC energy storage elements. The main DC energy storage element is divided into several local and small DC energy storage elements. This approach has the advantages of reduced switching loss and elimination of clamp circuit. The peak voltages applied to the semiconductor devices are clamped to local capacitors. The output performance is also better than the classical solution. The main disadvantage is the number of semiconductors and passive elements. Major difficulties in achieving the required control is related to monitoring of DC voltages across the capacitors. Each DC voltage in the capacitors has to be controlled via feedback. Then the control strategies are very complex. Furthermore, this converter can provide the buck-boost control of output voltage amplitude and can operate with arbitrary power factors.

The next means of achieving converter hybridisation is by using an additional small-scale DC circuit in the MC topologies. There are two types of matrix converter configurations, the single-stage (classical MC) and the two-stage (IMC); therefore, two ways to implant a hybrid MC are possible with MC and IMC.

The first solution is to connect an H-bridge inverter in series with each of the MC outputs as suggested in Fig. 2.65 [77]. The topology of the hybrid converter from Fig. 2.65 enables step-up or step-down voltage control. This solution has a serious disadvantage related to the high number of power semiconductors and DC link capacitors, which have to smooth down the power ripple (twice the output frequency) that is characteristic of a single-phase inverter [77]. Moreover, these converter topologies again have energy storage elements (e.g., electrolytic capacitor) which reduce their life time.

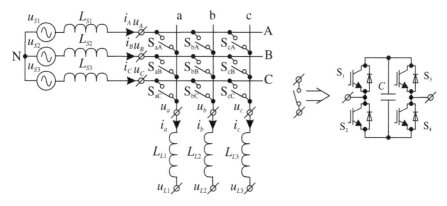

Fig. 2.64 Modular matrix converter (MMC) [7]

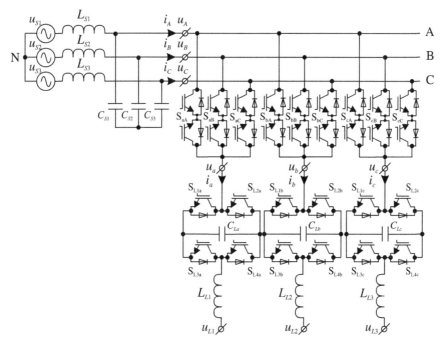

Fig. 2.65 Hybrid frequency converter with series connected MC and small scale H-bridge inverter on each output phase

The second solution is related to IMCs and solves the two most important drawbacks of the IMC—voltage transfer ratio equal to 0.866. It consists in introducing an auxiliary voltage supply in the form of an H-bridge inverter in the intermediary link of the IMC, with the purpose of compensating the voltage deficit. In this way, an increase of output voltage is obtained. The structure of an hybrid IMC with H-bridge

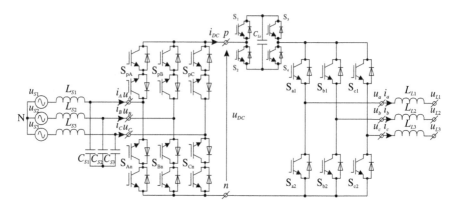

Fig. 2.66 Hybrid IMC with small scale H-bridge inverter in the intermediary link

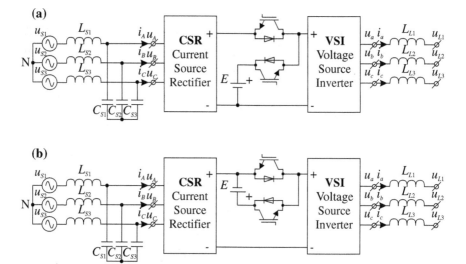

Fig. 2.67 Categories of hybrid IMC topologies employing: **a** an auxiliary high-voltage DC source; **b** low-voltage DC source

inverter is shown in Fig. 2.66 [73, 77, 78, 81, 144]. However, these converter topologies also have energy storage elements and their construction requires considerable work.

Another concept of hybrid IMC is proposed in [73], where hybrid structures are based on combined auxiliary voltage source in the intermediate DC link of the IMC. These converter topologies may be classified into two main groups depending on DC voltage level, as shown in Fig. 2.67. In the topology with an auxiliary high-voltage DC source, this source is used in parallel connection with a DC link, whereas in the topology with an auxiliary low-voltage DC source, in series connection with a DC

(a)

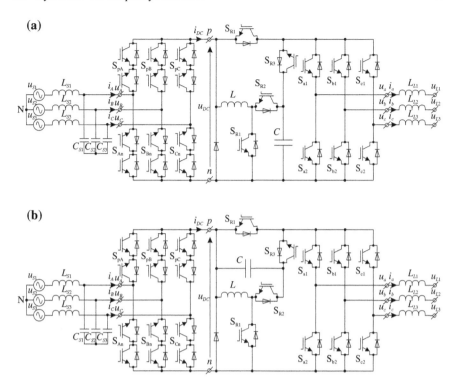

(b)

Fig. 2.68 Hybrid IMC structures: **a** with high-voltage auxiliary voltage source and reversible boost converter; **b** with low-voltage auxiliary voltage source and reversible flyback converter

link. The practical realisation of these concepts from Fig. 2.67, is shown in Fig. 2.68 [73, 144]. Unity voltage transfer is also obtained in these solutions.

In summary, hybrid concepts enable the enlargement of the voltage control range. To obtain these advantages, however, there is a higher complexity in the power stages and in their control. Furthermore, the DC energy storage element is still needed.

2.5 Summary of Topology Review

This section has dealt with a comprehensive review of AC–AC frequency converter topologies. This chapter has focused on inverter technologies with special attention to AC–AC conversion without DC energy storage elements. The potential converter topologies may be classified into three main groups depending AC–AC conversion (Fig. 2.2). Various inverter topologies have been presented, compared and evaluated against demands, lifetime, component ratings and voltage gain control capabilities.

Table 2.8 Summary of comparing frequency converter topologies

Name of converter	Modulation strategy	Voltage gain
With DC energy storage element		
Frequency converter with VSI	Space vector modulation (SVM)	>1
Frequency converter with CSI	Space vector modulation (SVM)	>1
Without DC energy storage element		
Voltage source matrix converter	Classical Venturini	0.5
	Improvement Venturini	0.866
	Scalar	0.866
	Indirect with fictitious DC link	1.05
	Space vector modulation (SVM)	0.866
	Duty-cycle SVM	1.155
Current source matrix converter	Classical Venturini	>1
	Space vector modulation (SVM)	>1
Indirect matrix converter (IMC)	Space vector modulation (SVM)	0.866
Sparse matrix converter (SMC)	Space vector modulation (SVM)	0.866
Very sparse matrix converter (VSMC)	Space vector modulation (SVM)	0.866
Ultra sparse matrix converter (USMC)	Space vector modulation (SVM)	0.866
Matrix-reactance frequency converters	Classical Venturini	>1
Cascaded connected MRC and MC	Space vector modulation (SVM)	>1
Multilevel indirect matrix converter	Space vector modulation (SVM)	>0.866
Multilevel direct matrix converter	Space vector modulation (SVM)	>0.8
Hybrid with small DC energy storage element		
Modular matrix converter	Space vector modulation (SVM)	>1
Hybrid direct MC with H bridge in output	Space vector modulation (SVM)	>1
Hybrid IMC with H bridge in DC link	Space vector modulation (SVM)	>1
Hybrid IMC with high-voltage auxiliary voltage source	Space vector modulation (SVM)	>1
Hybrid IMC with low-voltage auxiliary voltage source	Space vector modulation (SVM)	>1

The continual development of power electronic converters, for a range of applications, is characterised by the requirements for higher efficiency, lower volume, lower weight and lower production costs [82]. Frequency converters with DC energy storage have these desirable properties, but the voltage gain ratio is less than the one in most topologies and modulation strategies. For the topologies with modulation strategies with voltage gain ratio greater than one, there is low frequency deformation of output/input waveforms. Only on a few topologies without DC energy storage is buck-boost regulation of output voltages possible [8–10, 45, 46, 66, 79, 80, 87, 104, 153, 154]. Such topologies are not widely discussed in the literature. More interesting are topologies of matrix-reactance frequency converters, which will be analysed in detail in the following chapters. Finally, to complete this chapter, it is presented in Table 2.8, a summary comparison of the voltage gain ratio of frequency converter topologies.

References

1. Ahmed SM, Iqbal A, Abu-Rub H, Rodriguez J, Rojas CA, Saleh M (2011) Simple carrier-based PWM technique for a three-to-nine-phase direct AC–AC converter. IEEE Trans Ind Electron 58(11):5014–5023
2. Ahmed SM, Iqbal A, Abu-Rub H (2011) Generalized duty-ratio-based pulsewidth modulation technique for a three-to-k phase matrix converter. IEEE Trans Ind Electron 58(9):3925–3937
3. Alesina A, Venturini M (1989) Analysis and design of optimum-amplitude nine-switch direct AC-AC converters. IEEE Trans Power Electron 4(1):101–112
4. Alesina A, Venturini M (1988) Intrinsic amplitude limits and optimum design of 9-switches direct PWM AC-AC converters. In: Proceedings of IEEE power electronics specialists conference, PESC'88, Kyoto, Japan, pp 1284–1291
5. Andreu J, de Algeria I, Kortabarria I, Bidarte U, Ceballo S (2006) Matrix converter protection: active and passive strategy considerations. WSEAS Trans Power Syst 1(10):1698–1702
6. Andreu J, Kortabarria I, Ormaetxea E, Ibarra E, Martin JL, Apinaniz S (2012) A step forward towards the development of reliable matrix converters. IEEE Trans Ind Electron 59(1):167–183
7. Angkititrakul S, Erickson RW (2004) Control and implementation of a new modular matrix converter. In: Proceedings of IEEE applied power electronics conference and exposition, APEC'04, vol 2, Anaheim, US, pp 813–819
8. Antic D, Klaassens JB, Deleroi W (1993) A new power topology, suitable for low stator frequency operation of an induction machine. In: Proceedings of IEEE applied power electronics conference and exposition, APEC'93, San Diego, US, pp 146–152
9. Antic D, Klaassens JB, Deleroi W (1993) An integrated boost-buck and matrix converter topology for low speed drives. In: Proceedings of the EPE'93, Brighton, UK, pp 21–26
10. Antic D, Klaassens JB, Deleroi W (1994) Side effects in low-speed AC drives. In: Proceedings of IEEE power electronics specialists conference, PESC'94, Taipei, Taiwan, pp 998–1002
11. Arrilaga J, Watson NR (2003) Power system harmonics, 2nd edn. Wiley, Chichester
12. Arevalo SL, Zanchetta P, Wheeler PW, Trentin A, Empringham L (2010) Control and implementation of a matrix-converter-based AC ground power-supply unit for aircraft servicing. IEEE Trans Ind Electron 57(6):2076–2084
13. Beasant RR, Beatie WC, Refsum A (1990) An approach to the realisation of a high power venturini converter. In: IEEE industrial electronics, pp 291–297
14. Bernet S, Bernet K, Lipo TA (1996) The auxiliary resonant commutated pole matrix converter—a new topology for high power applications. In: Proceedings of IEEE industry applications conference annual meeting, IAS'96, vol 2, San Diego, US, pp 1242–1249
15. Bhangu BS, Snary P, Bingham CM, Stone DA (2005) Sensorless control of deep-sea ROVs PMSMs excited by matrix converter. In: Proceedings of the European conference on power electronics and applications, EPE 2005, Dresden, Germany (CD-ROM)
16. Bhowmik S, Spée R (1993) A guide to the application-oriented selection of AC/AC converter topologies. IEEE Trans Power Electron 8(2):156–163
17. Blaabjerg F, Casadei D, Klumpner C, Matteini M (2002) Comparison of two current modulation strategies for matrix converters under unbalanced input voltage conditions. IEEE Trans Ind Electron 49(2):289–296
18. Bland MJ, Clare JC, Wheeler PW, Empringham L, Apap M (2004) An auxiliary resonant soft switching matrix converter. In: Proceedings of IEEE power electronics specialists conference, PESC'04, vol 3, Aachen, Germany, pp 2393–2399
19. Bucknall RWG, Ciaramella KM (2010) On the conceptual design and performance of a matrix converter for marine electric propulsion. IEEE Trans Power Electron 25(6):1497–1508
20. Burany N (1989) Safe control of four-quadrant switches. In: Conference record of the IEEE industry applications conference annual meeting, IAS'89, pp 1190–1194
21. Casadei D (2005) Tutorial on matrix converters. In: Proceedings of power electronics and intelligent control for energy conservation conference, PELINCEC'05, Warsaw, Poland

22. Casadei D, Grandi G, Serra G, Tanti A (1993) Space vector control of matrix converters with unity input power factor and sinusoidal input/output waveforms. In: Proceedings of European conference on power electronics and applications, EPE'93, vol 7, Brighton, UK, pp 170–175
23. Casadei D, Serra G, Tani A (1998) Reduction of the input current harmonic content in matrix converters under input/output unbalance. IEEE Trans Ind Electron 45(3):401–411
24. Casadei D, Serra G, Tani A, Nielsen P (1995) Performance of SVM controlled matrix converter with input and output unbalanced conditions. In: Proceedings of European conference on power electronics and applications, EPE'95, vol 2, Seville, Spain, pp 628–633
25. Casadei D, Serra G, Tani A, Zarri L (2006) A review on matrix converters. Przegląd Elektrotechniczny (Electr Rev) 2:15–25
26. Casadei D, Serra G, Tani A, Zarri L (2005) Experimental behavior of a matrix converter prototype based on new power modules. Automatika (J Cont Measure Electron Comput Commun) 46(1–2):83–91
27. Casadei D, Serra G, Tani A, Zarri L (2009) Optimal use of zero vectors for minimizing the output current distortion in matrix converters. IEEE Trans Ind Electron 56(2):326–336
28. Casadei D, Serra G, Tanti A, Zaroi L (2002) Matrix converter modulation strategies: a new general approach based on space-vector representation of switch state. IEEE Trans Ind Electron 49(2):370–381
29. Casadei D, Trentin A, Matteini M, Calvini M (2003) Matrix converter commutation strategy using both output current and input voltage sign measurement. In: Proceedings of European conference on power electronics and applications, EPE'03, Toulouse, France, pp P1–P10 (CD-ROM)
30. Cho JG, Cho GH (1991) Soft switched matrix converter for high frequency direct AC-to-AC power conversion. In: Proceedings of European conference on power electronics and applications, EPE'91, Florence, Italy, pp 4-196–4-201
31. Empringham L, de Lillo L, Khwan-On S, Brunson C, Wheeler PW, Clare JC (2011) Enabling technologies for matrix converters in aerospace applications. In: Proceedings of international conference—-workshop compatibility and power electronics, CPE'2011, Tallinn, Estonia, pp 451–456
32. Empringham L, Wheeler PW, Clare JC (1998) Bi-directional switch current commutation for matrix converter applications. In: Proceedings of PE matrix converter, Prague, Czech Republic, pp 42–47
33. Empringham L, Wheeler PW, Clare JC (1998) Intelligent commutation of matrix converter bi-directional switch cells using novel gate drive techniques. In: Proceedings of power electronics specialists conference, PESC'98, Fukuoka, Japan, pp 707–713
34. Enjeti PN, Ziogas PD, Lindsay JF (1991) A current source PWM inverter with instantaneous current control capability. IEEE Trans Ind Appl 27:582–588
35. Erickson RW, Al-Naseem OA (2001) A new family of matrix converters. In: Proceedings of IEEE industrial electronics society conference, IECON'01, vol 2, Denver, US, pp 1515–1520
36. Fedyczak Z, Szcześniak P (2006) Koncepcja matrycowo-reaktancyjnego przemiennika częstotliwości typu Ćuk (in Polish). Przegląd Elektrotechniczny (Electr Rev) 7(8):42–47
37. Fedyczak Z, Szcześniak P (2006) Koncepcja matrycowo-reaktancyjnego przemiennika częstotliwości typu Zeta (in Polish). Wiadomości Elektrotechniczne (Electrotech News) 3:26–29
38. Fedyczak Z, Szcześniak P (2012) Matrix-reactance frequency converters using an low frequency transfer matrix modulation method. Electr Power Syst Res 83(1):91–103
39. Fedyczak F, Szcześniak P (2009) Modelling and analysis of matrix-reactance frequency converters using voltage source matrix converter and LF transfer matrix modulation method. Przegląd Elektrotechniczny (Electr Rev) 2:125–130
40. Fedyczak Z, Szcześniak P (2007) New matrix-reactance frequency converters—conception description. In: Orłowska-Kowalska T (ed) Power electronics and electrical drives: selected problems. Wrocław Technical University Press, Wrocław, pp 71–84
41. Fedyczak Z, Szcześniak P (2005) Study of matrix-reactance frequency converter with buck-boost topology. In: Proceedings of power electronics and intelligent control for energy conservation conference, PELINCEC'05, Warsaw, Poland (CD-ROM)

42. Fedyczak Z, Szcześniak P, Jankowski M (2005) Koncepcja matrycowo-reaktancyjnego przemiennika częstotliwości typu buck-bost (in Polish). Sterowanie w Energoelektronice i Napędzie Elektrycznym, SENE'05, number 1, Łódź, Poland, pp 101–106

43. Fedyczak Z, Szcześniak P, Kaniweski J (2007) Direct PWM AC choppers and frequency converters. In: Korbicz J (ed) Measurements models systems and design. Transport and Communication Publishers, Warsaw, pp 393–424

44. Fedyczak Z, Szcześniak P, Klytta M (2006) Matrix-reactance frequency converter based on buck-boost topology. In: Proceedings of power electronics and motion control conference, EPE-PEMC'06, Portoroz, Slovenia, pp 763–768

45. Fedyczak Z, Szcześniak P, Korotyeyev I (2008) Generation of matrix-reactance frequency converters based on unipolar PWM AC matrix-reactance choppers. In: Proceedings of IEEE power electronics specialists conference, PESC'08, Rhodes, Greece, pp 1821–1827

46. Fedyczak Z, Szcześniak P, Korotyeyev I (2008) New family of matrix-reactance frequency converters based on unipolar PWM AC matrix-reactance choppers. In: Proceedings of power electronics and motion control conference, EPE-PEMC'08, Poznań, Poland, pp 236–243

47. Fedyczak Z, Szcześniak P, Kaniweski J, Tadra G (2009) Implementation of three-phase frequency converters based on PWM AC matrix-reactance chopper with buck-boost topology. In: Proceedings of European conference on power electronics and applications, EPE'09, Barcelona, Spain, pp P1–P10 (CD-ROM)

48. Fedyczak Z, Tadra G, Klytta M (2010) Implementation of the current source matrix converter with space vector modulation. In: Proceedings of power electronics and motion control conference, EPE-PEMC'10, Ohrid, Macedonia (CD-ROM)

49. Fedyczak Z, Tadra G, Szczesniak P (2010) Three-phase AC systems interfaced by current source matrix converter with space vector modulation. In: International school on nonsinusoidal currents and compensation, ISNCC'2010, Łagów, Poland, pp 107–112

50. Fortescue CL (1918) Method of symmetrical coordinates applied to the solution of polyphase networks. Trans AIEE (part II) 37:1027–1140

51. Gyugi L, Pelly B (1976) Static power frequency changers: theory, performance and applications. Wiley, New York

52. Hava AM, Kerkman RJ, Lipo TA (1999) Simple analytical and graphical methods for carrier-based PWM-VSI drives. IEEE Trans Power Electron 14(1):49–61

53. He B, Wang X, Lin H, She H (2009) Research on two-step voltage-controlled commutation strategies for matrix converter. In: Proceedings of IEEE international power electronics and motion control conference, IPEMC '09, Wuhan, pp 1745–1751

54. Helle L, Munk-Nielsen S (2001) A novel loss reduced modulation strategy for matrix converters. In: Proceedings of IEEE power electronics specialists conference, PESC'01, vol 2, Vancouver, Canada, pp 1102–1107

55. Helle L, Larsen KB, Jorgensen HA, Munk-Nielsen S (2004) Evaluation of modulation schemes for three-phase to three-phase matrix converters. IEEE Trans Ind Electron 51(1):158–171

56. Herrero L, de Pablo S, Herrero LC, de Pablo S, Martin F, Ruiz JM, Gonzalez JM, Rey AB (2007) Comparative analysis of the techniques of current commutation in matrix converters. In: Proceedings of IEEE international symposium on industrial electronics, ISIE'07, pp 521–526

57. Hey HL, Pinheiro H, Pinheiro JR (1995) A new soft-switching AC-AC matrix converter, with a single actived commutation auxiliary circuit. In: Proceedings of IEEE power electronics specialists conference, PESC '95, vol 2, Atlanta, US, pp 965–970

58. Hofmann W, Ziegler M (2001) Multi-step commutation and control policies for matrix converters. In: Proceedings of international conference on power electronics, ISPE'01, Seoul, Korea, pp 795–802

59. Holmes DG, Lipo TA (2003) Pulse width modulation for power converters. Principle and practice. Wiley-IEEE, New York

60. Hornkamp M, Loddenkötter M, Muenzer M, Simon O, Bruckmann M (2001) EconoMAC the first all-in-one IGBT module for matrix converters. In: Proceedings of drives and controls and power electronics conference, London, UK, pp 35–39

61. Huber L, Borojević D (1995) Space vector modulated three-phase to three-phase matrix converter with input power factor correction. IEEE Trans Ind Appl 31(6):1234–1246
62. Huber L, Borojević D, Burany N (1989) Voltage space vector based PWM control of forced commutated cycloconverters. In: Proceedings of industrial electronics society annual conference, vol 1, IECON'89, pp 106–111
63. Iimori K, Shinohara K, Yamamoto K (2006) Study of dead time of PWM rectifier of voltage-source inverter without DC-link components and its operating characteristics of induction motor. IEEE Trans Ind Appl 42(2):518–525
64. Iimori K, Shinohara K, Tarumi O, Fu Z, Muroya M (1997) New current-controlled PWM rectifier voltage source inverter without DC-link components. In: Proceedings of power conversion conference, PCC'97, vol 2, Nagaoka, Japan, pp 783–786
65. Ishiguro A, Furuhashi T, Okuma S (1991) A novel control method for forced commutated cycloconverters using instantaneous values of input line-to-line voltages. IEEE Trans Ind Electron 38(3):166–172
66. Itoh J-I, Koiwa K, Kato K (2010) Input current stabilization control of a matrix converter with boost-up functionality. In: Proceedings of international power electronics conference, IPEC 2010, Sapporo, Japan
67. Jia S, Tseng KJ, Wang X (2005) Study on reverse recovery characteristics of reverse-blocking IGBT applied in matrix converter. In: Proceedings of IEEE applied power electronics conference and exposition, APEC'05, vol 3, Austin, US, pp 1917–1921
68. Jussila M, Tuusa H (2007) Comparison of simple control strategies of space-vector modulated indirect matrix converter under distorted supply voltage. IEEE Trans Power Electron 22(1):139–148
69. Jussila M, Salo M, Tuusa H (2003) Realization of a three-phase indirect matrix converter with an indirect vector modulation method. In: Proceedings of power electronics specialist conference, PESC'03, vol 2, Acapulco, Meksyk, pp 689–694
70. Kanaan HY, Al-Hadad K (2003) A new average modeling and control design applied to a nine-switch matrix converter with input power factor correction. In: Proceedings of EPE'03, Toulouse, France (CD-ROM)
71. Kato K, Itoh J-I (2007) Improvement of input current waveforms for a matrix converter using a novel hybrid commutation method. In: Proceedings of power conversion conference, PCC'07, Nagoya, Japan, pp 763–768
72. Kaźmierkowski MP, Krishnan R, Blaabjerg F (2002) Control in power electronics: selected problems. Academic Press Series in Engineering, New York
73. Klumpner C (2005) Hybrid direct power converters with increased/higher than unity voltage transfer ratio and improved robustness against voltage supply disturbances. In: Proceedings of power electronics specialists conference, PESC'05, pp 2383–2389
74. Klumpner C, Blaabjerg F (2002) Experimental evaluation of ride-through capabilities for a matrix converter under short power interruptions. IEEE Trans Ind Electron 49(2):315–324
75. Klumpner C, Blaabjerg F (2004) Short term braking capability during power interruptions for integrated matrix converter-motor drives. IEEE Trans Power Electron 2:303–311
76. Klumpner C, Blaabjerg F (2003) Two-stage direct power converters: an alternative to matrix converters. In: IEE matrix converter seminar, Birmingham, UK
77. Klumpner C, Pitic C (2008) Hybrid matrix converter topologies: an exploration of benefits. In: Proceedings of power electronics specialists conference, PESC'08, Rhodes, Greece, pp 2–8
78. Klumpner C, Wijekoon T, Wheeler P (2005) A new class of hybrid AC/AC direct power converters. In: Proceedings of IAS annual meeting industry applications conference, IAS'05, vol 4, Hong Kong, pp 2374–2381
79. Koiwa K, Itoh J-I (2011) A gain design method of a damping control for a matrix converter. 2011 annual meeting IEEJ, Toyonaka-city, Osaka, Japan, pp 1–2
80. Koiwa K, Itoh J-I (2011) Experimental verification for a matrix converter with a V-connection AC chopper. In: Proceedings of European conference on power electronics and applications, EPE'11, Birmingham, UK, pp 1–10

81. Kolar JW, Baumann M, Schafmeister F, Ertl H (2002) Novel three-phase AC-DC-AC sparse matrix converter. In: Proceedings of IEEE applied power electronics conference and exposition, APEC'02, vol 2, Dallas, US, pp 777–791
82. Kolar JW, Drofenik U, Biela J, Heldwein M, Ertl H, Friedli T, Round SD (2008) PWM converter power density barriers. IEEJ Trans Ind Appl 128:468–480
83. Kolar JW, Friedli T, Krismer F, Round SD (2008) The essence of three-phase AC/AC converter systems. In: Proceedings of power electronics and motion control conference, EPE-PEMC'08, Poznań, Poland, pp 27–42
84. Kolar JW, Friedli T, Rodriguez J, Wheeler PW (2011) Review of three-phase PWM AC–AC converter topologies. IEEE Trans Ind Electron 58(11):4988–5006
85. Korotyeyev I, Fedyczak Z (2008) Steady and transient states modelling methods of matrix-reactance frequency converter with buck-boost topology. COMPEL (Int J Comput Math Electr Electron Eng) 28(3):626–638
86. Korotyeyev I, Fedyczak Z, Szcześniak P (2008) Steady and transient state analysis of a matrix-reactance frequency converter based on a boost PWM AC matrix-reactance chopper. In: Proceedings of the international school on nonsinusoidal currents and compensation, ISNCC'08, Łagów, Poland (CD-ROM)
87. Kwon WH, Cho GH (1993) Analyses of static and dynamic characteristics of practical step-up nine-switch convertor. IEE Proc-B 140(2):139–145
88. Kwon WH, Cho GH (1991) Analysis of non-ideal step down matrix converter based on circuit DQ transformation. In: Proceedings of power electronics specialists conference, PESC'91, Cambridge, US, pp 825–829
89. Lee MY, Klumpner C, Wheeler PW (2008) Experimental evaluation of the indirect three-level sparse matrix converter. In: Proceedings of IET international conference on power electronics, machines and drives, PEMD'08, York, UK, pp 50–54
90. Lee MY, Wheeler PW, Klumpner C (2007) Modulation method for the three-level-output-stage matrix converter under balanced and unbalanced supply condition. In: Proceedings of European conference on power electronics and applications, EPE'07, Alborg, Denmark, pp 1–10 (CD-ROM)
91. Lie X, Clare JC, Wheeler PW, Empringham L (2008) Space vector modulation for a capacitor clamped multi-level matrix converter. In: Proceedings of power electronics and motion control conference, EPE-PEMC'08, Poznań, Poland, pp 229–235
92. Loh PC, Blaabjerg F, Gao F, Baby A, Tan DA (2008) Pulsewidth modulation of neutral-point-clamped indirect matrix converter. IEEE Trans Ind Appl 44(6):1805–1814
93. Lutz J, Schlangenotto H, Scheuerman U, De Doncker R (2011) Semiconductor power devices. Physics, characteristics, reliability. Springer, Berlin
94. Mahlein J, Braun M (2000) A matrix converter without diode clamped over-voltage protection. In: Proceedings of international power electronics and motion control conference, IPEMC 2000, vol 2, Beijing, China, pp 817–822
95. Mahlein J, Bruckmann M, Braun M (2002) Passive protection strategy for a drive system with a matrix converter and an induction machine. IEEE Trans Ind Electron 49(2):297–303
96. Mahlein J, Igney J, Braun M, Simon O (2001) Robust matrix converter commutation without explicit sign measurement. In: Proceedings of European conference on power electronics and applications, EPE'01 (CD-ROM)
97. Mahlein J, Igney J, Weigold J, Braun M, Simon O (2002) Matrix converter commutation strategies with and without explicit input voltage sign measurement. IEEE Trans Ind Electron 49(2):407–414
98. Majumdar G (2004) Future of power semiconductors. In: Proceedings of power electronics specialists conference, PESC'04, vol 1, Aachen, Germany, pp 10–15
99. Monteiro J, Silva JF, Pinto SF, Palma J (2009) Direct power control of matrix converter based unified power flow controllers. In: Proceedings of IEEE industrial electronics conference, IECON'09, Porto, Portugal, pp 1525–1530
100. Motto ER, Donlon JF, Tabata M, Takahashi H, Yu Y, Majumdar G (2004) Application characteristics of an experimental RB-IGBT (reverse blocking IGBT) module. In: Annual meeting of industry applications conference, IAS'04, vol 3, pp 1540–1544

101. Nielsen P, Blaabjerg F, Pedersen JK (1999) New protection issues of a matrix converter: design considerations for adjustable-speed drives. IEEE Trans Ind Appl 35(5):1150–1161
102. Nielsen P, Blaabjerg F, Pedersen JK (1996) Space vector modulated matrix converter with minimized number of switchings and a feedforward compensation of input voltage unbalance. In: Proceedings of international power electronics, drives and energy systems for industrial, growth, PEDES'96, vol 2, pp 833–839
103. Nikkhajoei H (2007) A current source matrix converter for high-power applications. In: Proceedings of IEEE power electronics specialists conference, PESC'07, Orlando, US, pp 2516–2521
104. Obuchov AY, Otchenasch W, Zinoviev GS (2000) Buck-boost AC-AC voltage controllers. In: Proceedings of international conference on power electronics and motion control, EPE-PEMC 2000, Košice, Slovakia, pp 2.194–2.197
105. Ormaetxea E, Andreu J, Kortabarria I, Bidarte U, Martinez de Alegria I, Ibarra E, Olaguenaga E (2011) Matrix converter protection and computational capabilities based on a system on chip design with an FPGA. IEEE Trans Power Electron 26(1):272–287
106. Oyama J, Xia X, Higuchi T, Yamada E (1997) Displacement angle control of matrix converter. In: Proceedings of IEEE power electronics specialists conference, PESC'97, St. Louise, US, pp 1033–1039
107. Pan CT, Chen TC, Shieh JJ (1993) A zero switching loss matrix converter. In: Proceedings of power electronics specialists conference, PESC'93, Seattle, US, pp 545–550
108. Pinto FS, Silva FJ (1999) Sliding mode control of space vector modulated matrix converter with sinusoidal input/output waveforms and near unity input power factor. In: Proceedings of European conference on power electronics and applications, EPE'99, Lausanne, Switzerland, pp 1–9
109. Rodriguez J (1983) A new control technique for AC-AC converters. In: Proceedings of control in power electronics and electrical drives conference, IFAC'83, Lausanne, Switzerland, pp 203–208
110. Rodriguez J, Rivera M, Kolar JW, Wheeler PW (2012) A review of control and modulation methods for matrix converters. IEEE Trans Ind Electron 59(1):58–70
111. Roy G, April GE (1989) Cycloconverter operation under a new scalar control algorithm. In: Proceedings of power electronics specialists conference, PESC'89, vol 1, Milwaukee, US, pp 368–375
112. Roy G, April GE (1991) Direct frequency changer operation under a new scalar control algorithm. IEEE Trans Power Electron 6(1):100–107
113. Roy G, Duguay L, Manias S, April GE (1987) Asynchronous operation of cycloconverter with improved voltage gain by employing a scalar control algorithm. In: Proceedings of IEEE-IAS annual meeting, pp 889–898
114. Rząsa J (2007) Wielopoziomowy przekształtnik matrycowy sterowany metodą venturiniego (in Polish). Przegląd Elektrotechniczny (Electr Rev) 2:57–64
115. Satish T, Mohapatra KK, Mohan N (2006) Modulation methods based on a novel carrier-based PWM scheme for matrix converter operation under unbalanced input voltages. In: Proceedings of applied power electronics conference and exposition, APEC'06, pp 127–132
116. Schafmeister F, Baumann M, Kolar JW (2002) Analytically closed calculation of the conduction and switching losses of three-phase AC-AC sparse matrix converters. In: Proceedings of international power electronics and motion control conference, EPE-PEMC'02, Dubrovnik, Croatia, pp 1–13 (CD-ROM)
117. Schonberger J, Friedli T, Round SD, Kolar JW (2007) An ultra sparse matrix converter with a novel active clamp circuit. In: Proceedings of power conversion conference, PCC'07, Nagoya, Japan, pp 784–791
118. She H, Lin H, He B, Wang X, Yue L, An X (2012) Implementation of voltage-based commutation in space-vector-modulated matrix converter. IEEE Trans Ind Electron 59(1):154–166
119. She H, Lin H, Wang X, Yue L, An X, He B (2011) Nonlinear compensation method for output performance improvement of matrix converter. IEEE Trans Ind Electron 58(9):3988–3999

120. Shi Y, Yang X, He Q, Wang Z (2004) Research on a novel multilevel matrix converter. In: Proceedings of IEEE power electronics specialists conference, PESC'04, vol 3, Aachen, Germany, pp 2413–2419

121. Simon O, Braun M (2001) Theory of vector modulation for matrix converters. In: Proceedings of European conference on power electronics and applications, EPE'01, Graz, Austria

122. Simon O, Mahlein J, Muenzer MN, Bruckmarm M (2002) Modern solutions for industrial matrix-converter applications. IEEE Trans Ind Electron 2:401–406

123. Sun K, Zhou D, Huang L, Matsuse K, Sasagawa K (2007) A novel commutation method of matrix converter fed induction motor drive using RB-IGBT. IEEE Trans Ind Appl 43(3):777–786

124. Svensson T, Alakula M (1991) The modulation and control of a matrix converter synchronous machine drive. In: Proceedings of European conference on power electronics and applications, EPE'91, Florence, Italy, pp 469–476

125. Szcześniak P (2010) Analiza i badania właściwości układu napędowego z matrycowo reaktancyjnym przemiennikiem częstotliwości o modulacji Venturiniego (in Polish). Przegląd Elektrotechniczny (Electr Rev) 6:155–158

126. Szcześniak P (2009) Analysis and testing matrix-reactance frequency converters. PhD thesis (in Polish), University of Zielona Góra, Zielona Góra

127. Szcześniak P (2007) Basic properties comparative study of matrix-reactance frequency converter based on buck-boost topology with Venturini control strategies. In: Proceedings of compatibility in power electronics, CPE'07, Gdańsk, Poland (CD-ROM)

128. Szcześniak P (2010) Modele matematyczne trójfazowych przemienników częstotliwości prądu przemiennego bazujących na topologii sterownika matrycowo-reaktancyjnego typu buck-boost (in Polish). Przegląd Elektrotechniczny (Electr Rev) 2:384–389

129. Szcześniak P, Fedyczak Z, Klytta M (2008) Modelling and analysis of a matrix-reactance frequency converter based on buck-boost topology by DQ0 transformation. In: Proceedings of power electronics and motion control conference, EPE-PEMC'08, Poznań, Poland, pp 165–172

130. Takei M, Naito T, Ueno K (2003) The reverse blocking IGBT for matrix converter with ultra-thin wafer technology. In: Proceedings of IEEE international symposium on power semiconductor devices and ICs, ISPSD'03, pp 156–159

131. Teichmann R, Oyama J (2002) ARCP soft-switching technique in matrix converters. IEEE Trans Ind Electron 49(2):353–361

132. Van Der Broeck H, Skudelny H, Stanke G (1986) Analysis and realization of a pulse width modulator based on voltage space vectors. In: Proceedings of IEEE-IAS'86, pp 244–251

133. Vargas R, Ammann U, Rodriguez J, Pontt J (2008) Predictive strategy to control common-mode voltage in loads fed by matrix converters. IEEE Trans Ind Electron 55(12):4372–4380

134. Vargas R, Ammann U, Hudoffsky B, Rodriguez J, Wheeler P (2010) Predictive torque control of an induction machine fed by a matrix converter with reactive input power control. IEEE Trans Power Electron 25(6):1426–1438

135. Venturini M, Alesina A (1980) The generalized transformer: a new bi-directional sinusoidal waveform frequency converter with continuously adjustable input power factor. In: Proceedings of IEEE power electronics specialists conference, PESC'80, pp 242–252

136. Villaça MV, Perin AA (1995) A soft switched direct frequency changer. In: Proceedings of IEEE industry applications conference, IAS'95, pp 2321–2326

137. Wang Y, Lu Z, Wen H, Wang Y (2005) Dead-time compensation based on the improved space vector modulation strategy for matrix converter. In: Proceedings of IEEE power electronics specialists conference, PESC'05, Recife, Brazil, pp 27–30

138. Wei L, Lipo TA (2001) A novel matrix converter topology with simple commutation. In: Proceedings of IEEE industry applications society annual meeting, IAS'01, vol 3, Chicago, US, pp 1749–1754

139. Wheeler PW, Clare J, Empringham L (2004) Enhancement of matrix converter output waveform quality using minimized commutation times. IEEE Trans Ind Electron 51(1):240–244

140. Wheeler PW, Clare J, Empringharn L, Bland M, Apap M (2002) Gate drive level intelligence and current sensing for matrix converter current commutation. IEEE Trans Ind Electron 49(2):382–389

141. Wheeler PW, Empringham L, Apap M, de Lilo L, Clare JC, Bradley K, Whitley C (2003) A matrix converter motor drive for an aircraft actuation system. In: Proceedings of the European conference on power electronics and applications, EPE'03, Toulouse, France (CD-ROM)

142. Wheeler PW, Rodriguez J, Clare JC, Empringham L, Weinstejn A (2002) Matrix converters: a technology review. IEEE Trans Ind Electron 49(2):276–288

143. Wheeler PW, Lie X, Lee MY, Empringham L, Klumpner C, Clare J (2008) A review of multi-level matrix converter topologies. In: Proceedings of IET international conference on power electronics, machines and drives, PEMD'08, York, UK, pp 286–290

144. Wijekoon T, Klumper C, Zanchetta P, Wheeler PW (2008) Implementation of a hybrid AC-AC direct power converter with unity voltage transfer. IEEE Trans Power Electron 23(4):1918–1926

145. Wilson J (1978) The forced-commutated inverter as a regenerative rectifier. IEEE Trans Ind Appl 14:335–340

146. Xu L, Clare JC, Wheeler PW, Empringham L, Li Y (2012) Capacitor clamped multilevel matrix converter space vector modulation. IEEE Trans Ind Electron 59(1):105–115

147. Yang X, Shi Y, He Q, Wang Z (2004) A novel multi-level matrix converter. In: Proceedings of IEEE applied power electronics conference and exposition, APEC'04, vol 2, Anaheim, US, pp 832–835

148. Yoon Y-D, Sul S-K (2006) Carrier-based modulation technique for matrix converter. IEEE Trans Power Electron 21(6):1691–1703

149. Zhou D, Sun K, Liu Z, Huang L, Matsuse K, Sasagawa K (2007) A novel driving and protection circuit for reverse-blocking IGBT used in matrix converter. IEEE Trans Ind Appl 43(1):3–13

150. Ziegler M, Hofmann W (2000) A new two steps commutation policy for low cost matrix converters. In: Proceedings of PCIM conference, Nürnberg, Germany

151. Ziegler M, Hofmann W (2001) New one-step commutation strategies in matrix converters. In: Proceedings of power electronics and drive systems conference, PEDS'01, vol 2, Bali, Indonesia, pp 560–564

152. Ziegler M, Hofmann W (1998) Semi natural two steps commutation strategy for matrix converters. In: Proceedings of power electronics specialists conference, PESC'98, Fukuoka, Japan, pp 727–731

153. Zinoviev GS, Obuchov AY, Otchenasch WA, Popov WI (2000) Transformerless PWM AC boost and buck-boost converters (in Russian). Technicznaja Elektrodinamika 2:36–39

154. Zinoviev GS, Ganin M, Levin E, Obuchov AY, Popov V (2000) New class of buck-boost AC-AC frequency converters and voltage controllers. In: Proceedings of Korea-Russia international symposium on science and technology, KORUS'2000, Ulsan, Korea, pp 303–308

155. Ziogas PD, Khan SI, Rashid MH (1986) Analysis and design of forced commutated cyclo-converter structures with improved transfer characteristics. IEEE Trans Ind Electron IE-33: 271–280

156. Ziogas PD, Khan SI, Rashid MH (1985) Some improved forced commutated cycloconverters structures. IEEE Trans Ind Appl 1A-21:1242–1253

Chapter 3
Concept of Matrix-Reactance Frequency Converters

3.1 Introduction

There has been some discussion to improve the voltage transfer ratio of the matrix converter, which is presented in Chap. 2. One of the proposed solutions is the matrix-reactance frequency converter (MRFC). The first study of MRFCs (with only one topology) was published in 1993 by Antic et al. in [3–5]. Later, in 2000, Zinoviev et al. continued the research on the same MRFC topology [28, 40, 41]. This converter was based on the buck-boost matrix-reactance chopper (MRC) [7, 14, 15] with source switches connected as in a matrix converter [38]. In 2003 Fedyczak published a paper [7], which shows that the idea of matrix-reactance choppers with source or load switches connected as in a matrix converter is technically possible for all unipolar PWM AC matrix-reactance choppers [7, 14, 15]. Another series of papers were published concerned with the MRFC based on the buck-boost MRC [11–13, 16–18, 21, 24, 35–37] and the novel three structures of MRFC based on Zeta [9, 12, 17], Ćuk [8, 12, 17] and boost [25] MRC topologies. In the topologies presented in these papers the source switches of MRFC with buck-boost and Zeta topology and load switches of MRFC with Ćuk topology are arranged as in the voltage source matrix converter. However, in the topology presented in [25] the output switches are arranged as in the current source matrix converter. Such an approach gives the possibility to obtain a load output voltage much greater than the input voltage. The conception of the MRFC was continuously developed by the authors of the latter papers. In the papers [19, 20, 35] the generation concept for a whole family of MRFCs based on unipolar PWM AC MRC was presented. The family of MRFCs contains nine topologies based on buck-boost, Ćuk, Zeta, SEPIC or boost MRC structures. Generally, MRFCs are divided into two groups. The first group contains MRFCs with switches arranged as in a voltage source matrix converter. In the second group are MRFCs with switches arranged as in a current source matrix converter. MRFC topologies have been intensively studied during the last few years, and the results have been widely published [8–13, 16–21, 24, 25, 34–37].

P. Szcześniak, *Three-Phase AC–AC Power Converters Based on Matrix
Converter Topology*, Power Systems, DOI: 10.1007/978-1-4471-4896-8_3,
© Springer-Verlag London 2013

3.2 Topology Generation

The topologies of the members of the MRFC family can be derived from the general structure shown in Fig. 3.1. The topologies of the MRFC are based on three-phase unipolar MRC structures [7, 14, 15]. Each unipolar MRC is composed of two synchronous-connected switch (SCS) sets. Generally, two stages occur in the steady-state operation of unipolar chopper converters during one switching cycle. The MRFC circuit is derived from the MRC circuit by implementing a matrix-connected switch set (MCS) instead of one of synchronous-connected switch sets [35]. Thereby, it is possible to use 28 state operations during one switching cycle. It gives the possibility of obtaining load voltage frequency change, as is the case in a matrix converter, and the possibility of obtaining a load output voltage much greater than the input voltage, as is the case in matrix-reactance choppers.

The matrix-connected switch set can be considered a generalised voltage source matrix converter (VSMC) and current source matrix converter (CSMC), both of which are shown in Fig. 3.2 [26, 27].

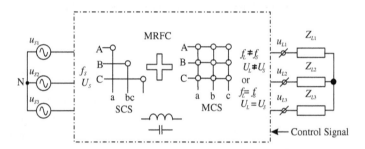

Fig. 3.1 General structure of basic matrix-reactance frequency converter, *SCS* synchronous-connected switches, *MCS* matrix-connected switches

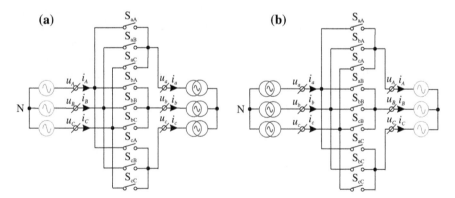

Fig. 3.2 Matrix-connected switch sets with, **a** voltage source matrix converter configuration, **b** current source matrix converter configuration

Fig. 3.3 Conceptual diagram
of matrix-reactance frequency
converter topology generation

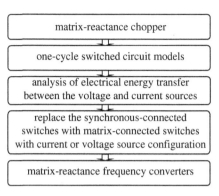

In the ideal case, the VSMC consists of a voltage source on the input side and current sources on the output side (Fig. 3.2a). In practice this can be realised by connecting capacitors and inductors on the input and output sides, respectively. The ideal CSMC consists of current sources on the input side and voltage sources on the output side (Fig. 3.2b). Similarly, as in the case of the VSMC, in practice a CSMC can be realised by connecting inductors and capacitors on the input and output sides, respectively.

No formal synthesis procedure is given for the derivation of the MRFC proposed in this monograph. The conceptual diagram on the basis of which the MRFC family is generated is shown in Fig. 3.3 [35]. The MRFC topologies are constructed on the basis of a unipolar MRC. The switched circuit models of the whole family of unipolar MRCs is shown in Fig. 3.4 [19, 20, 35].

In the first step, the one-cycle switched circuit models of all MRCs are constructed on the basis of these switched models. Since it is assumed that the two synchronous-connected switch sets are synchronised in their operation, the assumed switching sequence results in two modes of operation in one cycle time period. During the first mode the source synchronous-connected switches (SSCS) are turned on and the load synchronous-connected switches (LSCS) are turned off. During the second mode the switches are in the inverse states. Furthermore, in these models suitable voltage and current sources are taken into consideration instead of capacitors and inductors, respectively. The one-cycle switched circuit models of the whole family of MRCs shown in Fig. 3.4 for two modes, are shown in Fig. 3.5.

A new family of MRFCs has been developed based on the fundamental study of the one-cycle switched circuit models of unipolar MRCs. In the second step, the transfer of the energy between current and voltages sources, in one-cycle switched circuit models of unipolar MRC (Fig. 3.5), are analysed. As shown in Fig. 3.5, in each of the switch states, the synchronous-connected switch sets take part in two types of electrical energy transfer between the inner voltage and current sources. In the first one, the electrical energy is transferred through the turn-on switches from the voltage to the current sources. In the second one, the electrical energy is transferred through the turn-on switches from the current to the voltage sources.

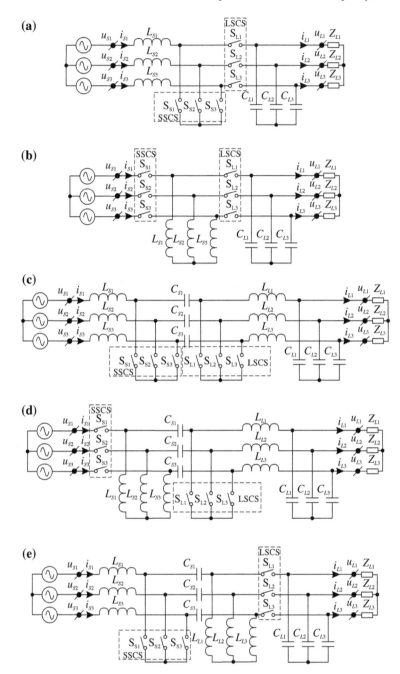

Fig. 3.4 Three-phase unipolar PWM AC MRC based on topology, **a** boost, **b** buck-boost, **c** Ćuk, **d** Zeta, **e** SEPIC; *SSCS* source synchronous-connected switches, *LSCS* load synchronous-connected switches

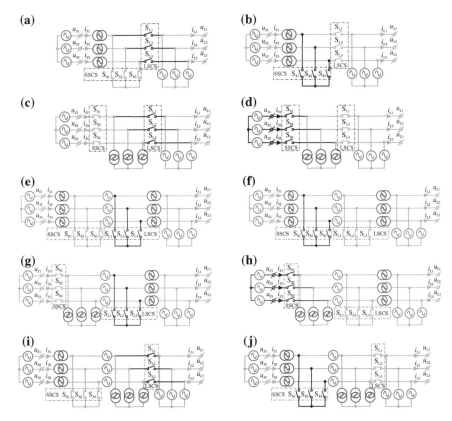

Fig. 3.5 One-cycle switched circuit models of three-phase unipolar PWM AC MRC based on topologies: **a, b** boost, **c, d** buck-boost, **e, f** Ćuk, **g, h** Zeta, **i, j** SEPIC; **a, c, e, g, i** for SSCS on and LSCS off, **b, d, f, h, j** for SSCS off and LSCS on

In the third step taking into account the kind of electrical energy transfer, one of synchronous-connected switch sets can be replaced by a matrix connected switch (MCS) set. If the electrical energy is transferred through the turn-on switches from the voltage to the current sources, then these switches can be replaced by the MCS set with voltage source matrix converter configuration (Fig. 3.2a). For the second kind of electrical energy transfer, when the electrical energy is transferred through the turn-on switches from the current to the voltage sources, the turn-on switches can be replaced by the MCS set with current source matrix converter configuration (Fig. 3.5b).

The circuit schemes of matrix-reactance frequency converters are given in Figs. 3.6, 3.7, 3.8, 3.9 and 3.10 [19, 20, 35]. On the basis of the presented procedure (Fig. 3.3), nine topologies of the MRFC were generated. The family of MRFCs contain two topologies based on buck-boost, Ćuk, SEPIC and Zeta matrix-reactance choppers and one topology based on the boost matrix-reactance chopper. The names and abbreviations of these converters are designated as follows [35]:

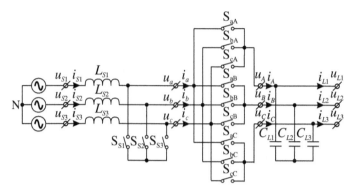

Fig. 3.6 Matrix-reactance frequency converter based on boost matrix reactance chopper (MRFC-b)

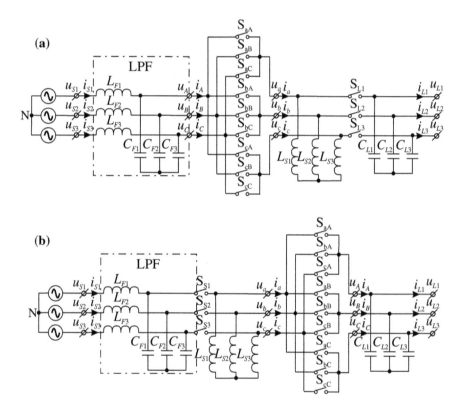

Fig. 3.7 Matrix-reactance frequency converters based on buck-boost matrix reactance chopper, **a** first topology (MRFC-I-b-b), **b** second topology (MRFC-II-b-b); *LPF* low pass filter

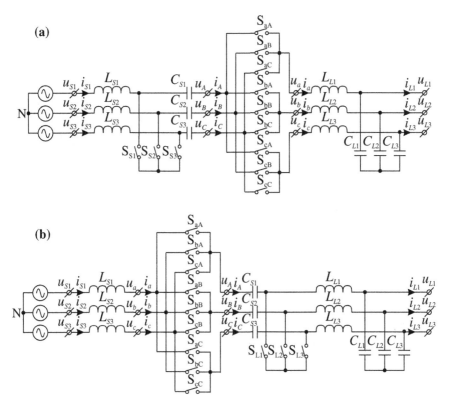

Fig. 3.8 Matrix-reactance frequency converters based on buck-boost and Ćuk matrix reactance chopper, **a** first topology (MRFC-I-c), **b** second topology (MRFC-II-c)

- MRFC-b—topology of MRFC based on boost MRC (Fig. 3.6),
- MRFC-I-b-b—first topology of MRFC based on buck-boost MRC (Fig. 3.7a),
- MRFC-II-b-b—second topology of MRFC based on buck-boost MRC (Fig. 3.7b),
- MRFC-I-c—first topology of MRFC based Ćuk MRC (Fig. 3.8a),
- MRFC-II-c—second topology of MRFC based on Ćuk MRC (Fig. 3.8b),
- MRFC-I-z—first topology of MRFC based on Zeta MRC (Fig. 3.9a),
- MRFC-II-z—second topology of MRFC based on Zeta MRC (Fig. 3.9b),
- MRFC-I-s—first topology of MRFC based on SEPIC MRC (Fig. 3.10a),
- MRFC-II-s—second topology of MRFC based on SEPIC MRC (Fig. 3.10b).

Furthermore, in both circuits of MRFCs based on buck-boost and Zeta topology the input low-pass filter L_F, C_F is used to reduce the source current deformation and elimination of the current spikes [33]. Adding an input filter to a converter will increase the complexity of the converter and has an influence on its functionality, stability, reliability size and cost, but is necessary to obtain sinusoidal input current waveforms.

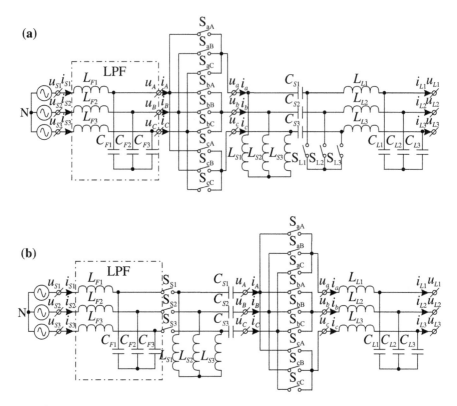

Fig. 3.9 Matrix-reactance frequency converters based on Zeta matrix reactance chopper, **a** first topology (MRFC-I-z), **b** second topology (MRFC-II-z); *LPF* low pass filter

All of the derived matrix-reactance frequency converters comprise standard matrix converter (with voltage or current source configuration) and the three-phase buck-boost matrix reactance chopper converters. The matrix converter part applies the pulse width modulation to produce a three-phase sinusoidal output voltage with the possibility to change load voltage (current) frequency. In addition, it enables the four-quadrant operation. A crucial fact is that all the generated MRFCs have the capability to obtain a load output voltage much greater than the input voltage, similar to the matrix-reactance chopper [7].

The derived MRFCs can be divided into two groups [19, 20, 35]. In the first group are converters which comprise voltage source matrix converter configuration (MRFC-I-b-b, MRFC-II-c, MRFC-I-z, MRFC-II-z). In the second group are converters witch current source matrix converter configuration (MRFC-II-b-b, MRFC-I-c, MRFC-I-s, MRFC-II-s, MRFC-b). A detailed analysis of selected converter topologies from both groups will be presented in the next section.

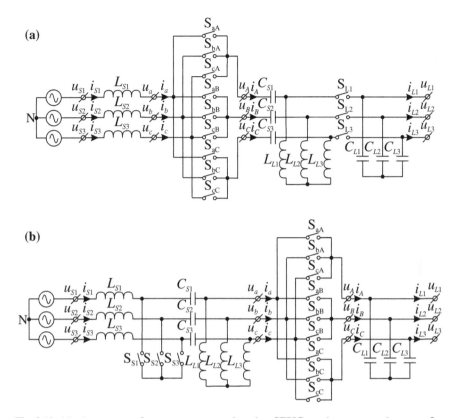

Fig. 3.10 Matrix-reactance frequency converters based on SEPIC matrix reactance chopper, **a** first topology (MRFC-I-s), **b** second topology (MRFC-II-s)

3.3 Topologies of Matrix-Reactance Frequency Converters with Voltage Source Matrix Converter

A voltage source matrix converter (VSMC) is an integral part of four matrix-reactance frequency converters: MRFC-I-b-b (Fig. 3.7b), MRFC-II-c (Fig. 3.8b), MRFC-I-z (Fig. 3.9a) and MRFC-II-z (Fig. 3.9b) [11, 35]. Figure 3.11 illustrates the description of the control strategy of the MRFC with VSMC, in general form [19, 20, 35]. Two stages occur (t_S and t_L) in the steady-state operation of the proposed converters over one switching cycle T_{Seq}.

The description of the control strategy of the MRFC-I-b-b and MRFC-I-z is presented in Fig. 3.11a, [35]. In each switching cycle T_{Seq}, in the interval t_S, the matrix connected switch sets are in the process of switching with selected switching modulation, while the load synchronous connected switch sets are turned off. The MCS output voltages u_a, u_b, u_c are formed by setting the requested output frequency f_L. In contrast, in the time period t_L all of the matrix connected switch sets are turned off

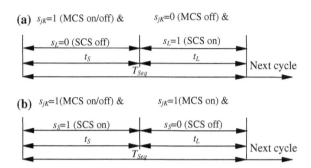

(a) $s_{jK}=1$ (MCS on/off) & $s_{jK}=0$ (MCS off) &

$s_L=0$ (SCS off) $s_L=1$ (SCS on)

t_S t_L

T'_{Seq} Next cycle

(b) $s_{jK}=1$(MCS on/off) & $s_{jK}=1$(MCS on) &

$s_S=1$ (SCS on) $s_S=0$ (SCS off)

t_S t_L

T_{Seq} Next cycle

Fig. 3.11 General form of the control strategy, for **a** MRFC-I-b-b and MRFC-I-z, and **b** MRFC-II-c and MRFC-II-z

and the load synchronous connected switch sets are turned on. The time interval t_L has an influence on the amplitude of load voltages u_{L1}, u_{L2}, u_{L3}. For the MRFC-II-c and MRFC-II-z the control strategy has a different form. In each switching cycle T_{Seq}, in the interval t_S the matrix connected switch sets switch with the selected switching modulation and simultaneously source synchronous connected switch sets are turned on. Similar to the MRFC-I-b-b and MRFC-I-z, the MCS output voltages u_a, u_b, u_c are formed by setting the requested output frequency f_L. In contrast, in the time period t_L all of the matrix connected switch sets are turned on and the source synchronous connected switch sets are turned off. Also, time t_L has an influence on the amplitude of load voltages u_{L1}, u_{L2}, u_{L3}.

The MRFC-I-b-b is analysed in detail, in order to analyse the MRFC with VSMC topology operations. Key idealised, theoretical waveforms concerning the operational stages in the MRFC-I-b-b converter are shown in Figs. 3.12 and 3.13 [35]. The modified, simplified, classical Venturini controlled strategy [38], which will be presented in the next subsection, is used for control of matrix connected switch sets. Figures 3.12 and 3.13 show the operating waveforms of the circuit for 1.5 periods of source phase voltage frequency, and zoom for two periods of switch sequence cycle T_{Seq}.

During the time period t_S, the output voltages of matrix connected switch sets u_a, u_b, u_c are synthesised by sequential piecewise sections of the input voltage waveforms u_A, u_B, u_C. The synthesis of output voltage u_a is shown in Fig. 3.12a–f. The duration of each piecewise section of the input voltage waveforms is controlled by the switch pulse duty factor s_{aA}, s_{aB}, s_{aC}, (Fig. 3.12a, b). The input voltages of matrix connected switch sets u_A, u_B, u_C are quasi-sinusoidal voltages of source filter capacitances C_{F1}, C_{F2}, C_{F3} (Fig. 3.12c, d). Similar to the output voltages, the input currents of matrix connected switch sets i_A, i_B, i_C are directly generated by the output currents i_a, i_b, i_c. The synthesis of input current i_A, is shown in Fig. 3.13a–f. The pieces of output current are selected on the basis of the pulse duty factor of switches s_{aA}, s_{bA}, s_{cA} (Fig. 3.13a, b). Source currents i_{S1}, i_{S2}, i_{S3} are filtered by a low pass LC filter. The performance of the source current is presented in Fig. 3.13c, d. In time t_S the output voltage shape waveforms are formed with setting frequency f_L.

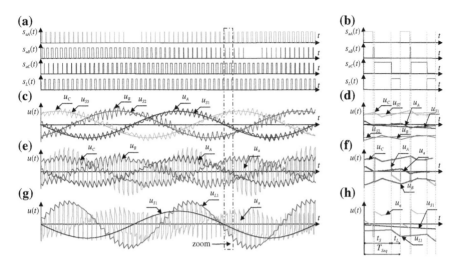

Fig. 3.12 Example of voltage time waveforms in MRFC-I-b-b for sequence frequency $f_{Seq} = 1\,\text{kHz}$ at pulse duty factor $D_S = 0.75$ and with load voltage setting frequency $f_L = 75\,\text{Hz}$

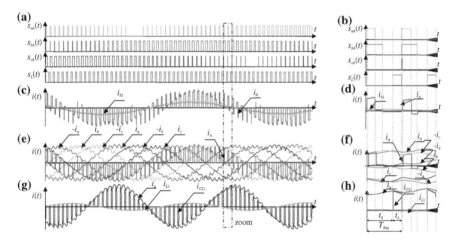

Fig. 3.13 Example of current time waveforms in MRFC-I-b-b for sequence frequency $f_{Seq} = 1\,\text{kHz}$ at pulse duty factor $D_S = 0.75$ and with load voltage setting frequency $f_L = 75\,\text{Hz}$

At the same time the electrical energy is stored in the inductor L_{S1}, L_{S2}, L_{S3}. The output voltages u_{L1}, u_{L2}, u_{L3} depend on the energy stored in load capacitors C_{L1}, C_{L2}, C_{L3}. Only energy stored in these capacitors is transferred to the load during the time period t_S.

During the time period t_L, the matrix connected switch sets are turned off and load synchronous switch sets are turned on. The energy stored in source inductors L_{S1}, L_{S2}, L_{S3} is transferred to the load capacitor C_{L1}, C_{L2}, C_{L3} and the loads are as shown in Fig. 3.13g, h, for example, for the capacitance current i_{CL1}. Then electrical energy is stored in the load capacitor C_{L1}, C_{L2}, C_{L3}. During this stage, instantaneous values of load voltages are equal to voltages u_a, u_b, u_c (Fig. 3.12g, h). It is possible to control the amplitude of output voltages u_{L1}, u_{L2}, u_{L3} through a change in the time interval t_L. As shown in Fig. 3.12g, the output voltage u_{L1} is greater than the source voltage u_{S1}.

3.4 Topologies of Matrix-Reactance Frequency Converters with a Current Source Matrix Converter

A current source matrix converter (CSMC) is an integral part of four matrix-reactance frequency converters: MRFC-b (Fig. 3.6) MRFC-II-b-b (Fig. 3.7b), MRFC-I-c (Fig. 3.8a), MRFC-I-s (Fig. 3.10a) and MRFC-II-s (Fig. 3.10b) [35]. In Fig. 3.14 a generalised control algorithm of the MRFC with CSMC is given. It consists of two parts t_S and t_L in each switching cycle T_{Seq}.

The description of the control strategy of the MRFC-b, MRFC-II-b-b and MRFC-II-s is given in Fig. 3.14a. In each switching cycle T_{Seq}, in the interval t_S the matrix connected switch sets are turned off and source synchronous connected switch sets are turned on. The electrical energy is stored in the inductor L_{S1}, L_{S2}, L_{S3}. In the time period t_L the matrix connected switch sets are switching with select switching modulation and simultaneously source synchronous connected switch sets are turned off. The energy stored in source inductors L_{S1}, L_{S2}, L_{S3} is transferred to the loads and at the same time the MCS output voltage shape waveforms u_A, u_B, u_C are formed by setting requested output frequency f_L. For the MRFC-I-c and MRFC-I-s the control strategy is given in Fig. 3.14b. In each switching cycle T_{Seq}, in the interval t_S the matrix connected switch sets are turned on, and load synchronous connected switch sets are turned off. The electrical energy is stored in the inductors L_{S1}, L_{S2}, L_{S3}. In the time period t_L the matrix connected switch sets switch with

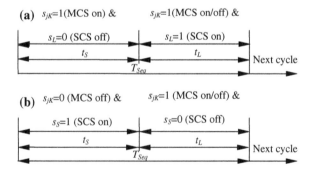

Fig. 3.14 General form of the control strategy for, **a** MRFC-b, MRFC-II-b-b and MRFC-II-s, and **b** MRFC-I-c and MRFC-I-s

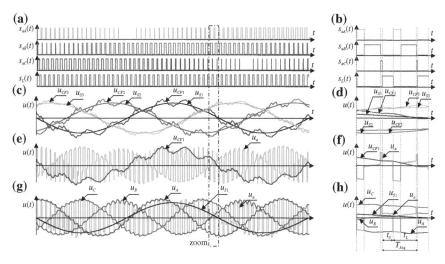

Fig. 3.15 Example of voltage time waveforms in MRFC-II-b-b for sequence frequency $f_{Seq} = 1\,kHz$ at pulse duty factor $D_S = 0.75$ and with load voltage setting frequency $f_L = 75\,Hz$

selected switching modulation and simultaneously the source synchronous connected switch sets are turned on. Similarly as in the MRFC-b, MRFC-II-b-b and MRFC-II-s, the energy stored in the source inductors L_{S1}, L_{S2}, L_{S3} is transferred to the loads and simultaneously the MCS output voltage shape waveforms u_A, u_B, u_C are formed by setting output frequency f_L. Studying the control strategies presented in Fig. 3.14 one can say that interval time t_S has an influence on the amplitude of load voltages u_{L1}, u_{L2}, u_{L3}.

The MRFC-II-b-b is analysed in detail, in order to analyse the MRFC with CSMC topology operations. Figures 3.15 and 3.16 illustrate the idealised, theoretical waveforms concerning the operational stages in the MRFC-II-b-b converter. Also, the modified, simplified, classical Venturini control strategy is used.

From Figs. 3.15 and 3.16, it can be observed that during the time period t_S, when the matrix connected switch sets are turned off and source synchronous switch sets are turned on, the electrical energy is stored in the inductors L_{S1}, L_{S2}, L_{S3}. The current of source switches is given in Fig. 3.16c, d. The input voltages of matrix connected switch sets u_a, u_b, u_c are equal to the quasi-sinusoidal voltages of source filter capacitances C_{F1}, C_{F2}, C_{F3} (Fig. 3.15e, f). The output voltages u_{L1}, u_{L2}, u_{L3} depend on the energy stored in load capacitors C_{L1}, C_{L2}, C_{L3}. Only energy stored in these capacitors is transferred to the load during the time period t_S (Fig. 3.16g, h).

During the time period t_L, the matrix connected switch sets are switched on and source synchronous switch sets are turned off (Figs. 3.15a, b, 3.16a, b). The energy stored in source inductors L_{S1}, L_{S2}, L_{S3} is transferred to the load capacitors C_{L1}, C_{L2}, C_{L3} and the loads. Simultaneously, the output currents i_A, i_B, i_C are synthesised by sequential piecewise sections of the input current waveforms i_a, i_b, i_c. The synthesis of output current i_A, is shown in Fig. 3.16e, f. The duration of each

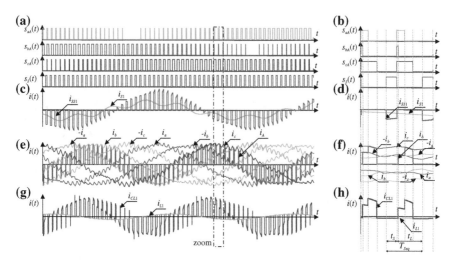

Fig. 3.16 Example of current time waveforms in MRFC-II-b-b for sequence frequency $f_{Seq} = 1\,kHz$ at pulse duty factor $D_S = 0.75$ and with load voltage setting frequency $f_L = 75\,Hz$

piecewise sections of the input current waveforms is controlled by the switch pulse duty factor s_{aA}, s_{bA}, s_{cA}, (Fig. 3.16a, b). At the same time the voltages u_a, u_b, u_c are synthesised from the output voltages u_A, u_B, u_C (Fig. 3.15g, h), controlled by the switch pulse duty factors s_{aA}, s_{aB}, s_{aC}, (Fig. 3.15a, b). In time t_L the output current shape waveforms are formed with setting frequency f_L. From the presented figures it can be observed that it is possible to control amplitude of output voltages u_{L1}, u_{L2}, u_{L3}. As is visible in Fig. 3.15g, the output voltage u_{L1} is greater than the source voltage u_{S1}.

3.5 Control Strategies

The general form of the control strategy descriptions of MRFC with VSMC are given in Fig. 3.11, whereas the control strategy descriptions of MRFC with CSMC are shown in Fig. 3.14. During each switching period, the switching pattern is divided into two parts. In one, the matrix connected switch sets are in the process of switching, whereas during the second part all the switch sets are either turned off or turned on. During one appropriate time period, matrix connected switch sets can be controlled with all the control strategies used in a matrix converter. From among different control techniques the best known are: classical Venturini control strategy [38], optimum (improved) Venturini control strategy [1, 2, 36], indirect modulation methods [42, 43], scalar modulation methods [30–32] and space vector modulation [6, 22, 23, 29].

This book considers classical, simplified Venturini control strategy for use in the control of matrix-reactance frequency converters. The proposed technique is based on classical Venturini modulation with only one part of the low frequency modulation matrix [29, 38, 39]. Taking into account only one basic solution defining $\omega_m = \omega_L - \omega_S$, and with limited switching time (t_S or t_L), the low frequency modulation matrix (3.1) for an MRFC based on VSMC and CSMC has been described by Eqs. (3.2) and (3.3), respectively [19, 20, 35].

$$\mathbf{M} = \begin{bmatrix} d_{aA} & d_{aB} & d_{aC} \\ d_{bA} & d_{bB} & d_{bC} \\ d_{cA} & d_{cB} & d_{cC} \end{bmatrix}, \tag{3.1}$$

$$d_{aA} = d_{bB} = d_{cC} = \frac{D_S}{3}(1 + 2q\cos(\omega_m t))$$

$$d_{aB} = d_{cA} = d_{bC} = \frac{D_S}{3}(1 + 2q\cos(\omega_m t - 2\pi/3)) \tag{3.2}$$

$$d_{aC} = d_{bA} = d_{cB} = \frac{D_S}{3}(1 + 2q\cos(\omega_m t - 4\pi/3)),$$

and

$$d_{aA} = d_{bB} = d_{cC} = \frac{(1 - D_S)}{3}(1 + 2q\cos(\omega_m t))$$

$$d_{aB} = d_{cA} = d_{bC} = \frac{(1 - D_S)}{3}(1 + 2q\cos(\omega_m t - 4\pi/3)) \tag{3.3}$$

$$d_{aC} = d_{bA} = d_{cB} = \frac{(1 - D_S)}{3}(1 + 2q\cos(\omega_m t - 2\pi/3)),$$

where q—setting voltage gain ($0 \leq q \leq 0.5$), D_S—sequence duty factor described by the following equation:

$$D_S = \frac{t_S}{T_{Sec}}. \tag{3.4}$$

The general forms of switching pattern for the classical simplified Venturini control strategy (3.1)–(3.4), for an MRFC with VSMC are given in Fig. 3.17, whereas for an MRFC with CSMC, they are shown in Fig. 3.18. A simplified realisation of the control strategies from Figs. 3.17 and 3.18 is depicted in Figs. 3.19 and 3.20, respectively [19, 20, 35]. The reference modulation wave is compared with a triangular carrier wave and the intersections define the switching instants.

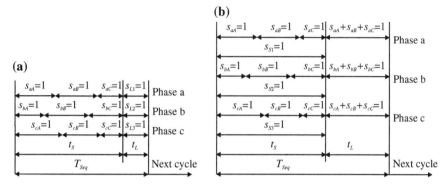

Fig. 3.17 The general form of switching pattern of MRFC with VSMC, **a** for MRFC-I-b-b and MRFC-I-z, **b** MRFC-II-c and MRFC-II-z

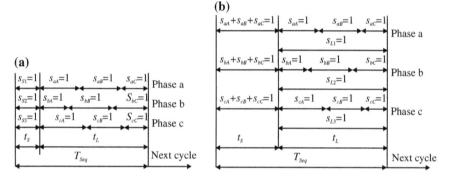

Fig. 3.18 The general form of switching pattern of MRFC with VSMC, **a** for MRFC-b, MRFC-II-b-b and MRFC-II-s, **b** MRFC-I-c and MRFC-I-s

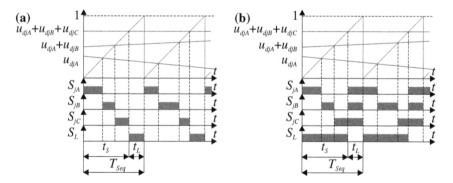

Fig. 3.19 A simplified realisation of the control strategies from Fig. 3.17: **a** for MRFC-I-b-b and MRFC-II-z, **b** MRFC-II-c and MRFC-II-z

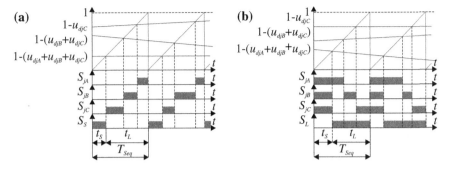

Fig. 3.20 A simplified realisation of the control strategies from Fig. 3.18: **a** for MRFC-b, MRFC-II-b-b and MRFC-II-s, **b** MRFC-I-c and MRFC-I-s

3.6 Chapter Summary

The description of a way to generate a novel family of matrix-reactance frequency converters (MRFC) has been presented in this chapter. These topologies are based on matrix-reactance choppers with source or load switches arranged as in matrix converter. In this way new topologies are constructed. Nine new topologies are suggested. The modification of Venturini control strategy is proposed for converter control. Furthermore, other modulation methods are indicated.

Matrix-reactance frequency topologies allow buck-boost output voltage regulation (similar to that in matrix-reactance choppers) and frequency change (similar to that in a matrix converter). In the next chapter a modelling approach and a study of the properties of MRFCs will be presented.

References

1. Alesina A, Venturini M (1989) Analysis and design of optimum-amplitude nine-switch direct AC-AC converters. IEEE Trans Power Electron 4(1):101–112
2. Alesina A, Venturini M (1988) Intrinsic amplitude limits and optimum design of 9-switches direct PWM AC-AC converters. In: Proceedings of IEEE power electronics specialists conference, PESC'88, Kyoto, Japan, pp 1284–1291
3. Antic D, Klaassens JB, Deleroi W (1993) A new power topology, suitable for low stator frequency operation of an induction machine. In: Proceedings of IEEE applied power electronics conference and exposition, APEC'93, San Diego, US, pp 146–152
4. Antic D, Klaassens JB, Deleroi W (1993) An integrated boost-buck and matrix converter topology for low speed drives. In: Proceedings of the EPE'93, Brighton, UK, pp 21–26
5. Antic D, Klaassens JB, Deleroi W (1994) Side effects in low-speed AC drives. In: Proceedings of IEEE power electronics specialists conference, PESC'94, Taipei, Taiwan, pp 998–1002
6. Casadei D (2005) Tutorial on matrix converters. In: Proceedings of power electronics and intelligent control for energy conservation conference, PELINCEC'05, Warsaw, Poland
7. Fedyczak Z (2003) PWM AC voltage transforming circuits (in Polish). Zielona Góra University Press, Zielona Góra

8. Fedyczak Z, Szcześniak P (2006) Koncepcja matrycowo-reaktancyjnego przemiennika częstotliwości typu Ćuk (In Polish). Przegląd Elektrotechniczny (Electr Rev) 7/8:42–47

9. Fedyczak Z, Szcześniak P (2006) Koncepcja matrycowo-reaktancyjnego przemiennika częstotliwości typu Zeta (in Polish). Wiadomości Elektrotechniczne (Electrotech News) 3: 26–29

10. Fedyczak Z, Szcześniak P (2012) Matrix-reactance frequency converters using an low frequency transfer matrix modulation method. Electr Power Syst Res 83(1):91–103

11. Fedyczak F, Szcześniak P (2009) Modelling and analysis of matrix-reactance frequency converters using voltage source matrix converter and LF transfer matrix modulation method. Przegląd Elektrotechniczny (Electr Rev) 2:125–130

12. Fedyczak Z, Szcześniak P (2007) New matrix-reactance frequency converters-conception description. In: Orłowska-Kowalska T (ed) Power electronics and electrical drives: selected problems. Wrocław Technical University Press, Wrocław, pp 71–84

13. Fedyczak Z, Szcześniak P (2005) Study of matrix-reactance frequency converter with buck-boost topology. In: Proceedings of power electronics and intelligent control for energy conservation conference, PELINCEC'05, Warsaw, Poland (CD-ROM)

14. Fedyczak Z, Klytta M, Strzelecki R. (2001) Three-phase AC/AC semiconductor transformer topologies and applications. In: Proceedings of power electronics devices compatibility conference, PEDC'01, Zielona Góra, Poland, pp 25–38

15. Fedyczak Z, Strzelecki R, Sozański K (2002) Review of three-phase AC/AC semiconductor transformer topologies and applications. In: Proceedings of symposium on power electronics, electrical drives, automation and motion, SPEEDAM'02, Ravello, Italy, pp B.5-19–B.5-24

16. Fedyczak Z, Szcześniak P, Jankowski M (2005) Koncepcja matrycowo-reaktancyjnego przemiennika częstotliwości typu buck-bost (in Polish). Sterowanie w Energoelektronice i Napędzie Elektrycznym, SENE'05, number 1, Łódź, Poland, pp 101–106

17. Fedyczak Z, Szcześniak P, Kaniweski J (2007) Direct PWM AC choppers and frequency converters. In: Korbicz J (ed) Measurements models systems and design. Transport and Communication Publishers, Warsaw, pp 393–424

18. Fedyczak Z, Szcześniak P, Klytta M (2006) Matrix-reactance frequency converter based on buck-boost topology. In: Proceedings of power electronics and motion control conference, EPE-PEMC'06, Portoroz, Slovenia, pp 763–768

19. Fedyczak Z, Szcześniak P, Korotyeyev I (2008) Generation of matrix-reactance frequency converters based on unipolar PWM AC matrix-reactance choppers. In: Proceedings of IEEE power electronics specialists conference, PESC'08, Rhodes, Greece, pp 1821–1827

20. Fedyczak Z, Szcześniak P, Korotyeyev I (2008) New family of matrix-reactance frequency converters based on unipolar PWM AC matrix-reactance choppers. In: Proceedings of power electronics and motion control conference, EPE-PEMC'08, Poznań, Poland, pp 236–243

21. Fedyczak Z, Szcześniak P, Kaniweski J, Tadra G (2009) Implementation of three-phase frequency converters based on PWM AC matrix-reactance chopper with buck-boost topology. In: Proceedings of European conference on power electronics and applications, EPE'09, Barcelona, Spain, pp P1–P10 (CD-ROM)

22. Huber L, Borojević D (1995) Space vector modulated three-phase to three-phase matrix converter with input power factor correction. IEEE Trans Ind Appl 31(6):1234–1246

23. Huber L, Borojević D, Burany N (1989) Voltage space vector based PWM control of forced commutated cycloconverters. In: Proceedings of industrial electronics society annual conference, IECON'89, vol 1, pp 106–111

24. Korotyeyev I, Fedyczak Z (2008) Steady and transient states modelling methods of matrix-reactance frequency converter with buck-boost topology. COMPEL (Int J Comput Math Electr Electron Eng) 28(3):626–638

25. Korotyeyev I, Fedyczak Z, Szcześniak P (2008) Steady and transient state analysis of a matrix-reactance frequency converter based on a boost PWM AC matrix-reactance chopper. In: Proceedings of the international school on nonsinusoidal currents and compensation, ISNCC'08, Łagów, Poland (CD-ROM)

26. Kwon WH, Cho GH (1993) Analyses of static and dynamic characteristics of practical step-up nine-switch convertor. IEE Proc-B 140(2):139–145
27. Kwon WH, Cho GH (1991) Analysis of non-ideal step down matrix converter based on circuit DQ transformation. In: Proceedings of power electronics specialists conference, PESC'91, Cambridge, US, pp 825–829
28. Obuchov AY, Otchenasch W, Zinoviev GS (2000) Buck-boost AC-AC voltage controllers. In: Proceedings of international conference on power electronics and motion control, EPE-PEMC 2000, Košice, Slovakia, pp 2.194–2.197
29. Rodriguez J, Rivera M, Kolar JW, Wheeler PW (2012) A review of control and modulation methods for matrix converters. IEEE Trans Ind Electron 59(1):58–70
30. Roy G, April GE (1989) Cycloconverter operation under a new scalar control algorithm. In: Proceedings of power electronics specialists conference, PESC'89, vol 1, Milwaukee, US, pp 368–375
31. Roy G, April GE (1991) Direct frequency changer operation under a new scalar control algorithm. IEEE Trans Power Electron 6(1):100–107
32. Roy G, Duguay L, Manias S, April GE (1987) Asynchronous operation of cycloconverter with improved voltage gain by employing a scalar control algorithm. In: Proceedings of IEEE-IAS annual meeting, pp 889–898
33. She H, Lin H, Wang X, Yue L (2009) Damped input filter design of matrix converter. In: Proceedings of international conference on power electronics and drive systems, PEDS'09, Taipei, Taiwan
34. Szcześniak P (2010) Analiza i badania właściwości układu napędowego z matrycowo reaktancyjnym przemiennikiem częstotliwości o modulacji Venturiniego (in Polish). Przegląd Elektrotechniczny (Electr Rev) 6:155–158
35. Szcześniak P (2009) Analysis and testing matrix-reactance frequency converters. PhD thesis (in Polish), University of Zielona Góra, Zielona Góra
36. Szcześniak P (2007) Basic properties comparative study of matrix-reactance frequency converter based on buck-boost topology with Venturini control strategies. In: Proceedings of compatibility in power electronics, CPE'07, Gdańsk, Poland (CD-ROM)
37. Szcześniak P, Fedyczak Z, Klytta M (2008) Modelling and analysis of a matrix-reactance frequency converter based on buck-boost topology by DQ0 transformation. In: Proceedings of power electronics and motion control conference, EPE-PEMC'08, Poznań, Poland, pp 165–172
38. Venturini M, Alesina A (1980) The generalized transformer: a new bi-directional sinusoidal waveform frequency converter with continuously adjustable input power factor. In: Proceedings of IEEE power electronics specialists conference, PESC'80, pp 242–252
39. Wheeler PW, Rodriguez J, Clare JC, Empringham L, Weinstejn A (2002) Matrix converters: a technology review. IEEE Trans Ind Electron 49(2):276–288
40. Zinoviev GS, Obuchov AY, Otchenasch WA, Popov WI (2000) Transformerless PWM AC boost and buck-boost converters (in Russian). Technicznaja Elektrodinamika 2:36–39
41. Zinoviev GS, Ganin M, Levin E, Obuchov AY, Popov V (2000) New class of buck-boost AC-AC frequency converters and voltage controllers. In: Proceedings of Korea-Russia international symposium on science and technology, KORUS'2000, Ulsan, Korea, pp 303–308
42. Ziogas PD, Khan SI, Rashid MH (1986) Analysis and design of forced commutated cycloconverter structures with improved transfer characteristics. IEEE Trans Ind Electron IE-33: 271–280
43. Ziogas PD, Khan SI, Rashid MH (1985) Some improved forced commutated cycloconverters structures. IEEE Trans Ind Appl 1A-21:1242–1253

Chapter 4
Modeling of Matrix-Reactance Frequency Converters

4.1 Introduction

In the past several years, there has been a lot of research done on power converters modelling aspect. One well-known approach to the modelling of PWM systems is to approximate their operation by averaging techniques. The generalised averaging method is based on the fact that the waveforms can be approximated using a defined time interval. This interval is determined by a switching period T_S or switching sequence period T_{Seq}.

In the literature there are given several average model derivations. For pulse width modulated (PWM) power converters, the most popular approaches are known as the state-space averaging method, the circuit averaging method and the switching functions-based model [2, 20, 23, 27]. Issues of frequency converter modelling have been addressed in a few papers in recent years. The papers [1, 3–19, 21, 22, 25, 26, 28–34] provide particularly valuable perspectives on these issues. There are several analytical methods for obtaining averaged modells. These include detailed circuit analysis using: four terminal network theory [33], signal flow graph [25, 26], graphical phasor [1, 29] and solving mathematical equations [5, 6, 14–19, 21, 22, 30, 32]. Analytical methods for averaged models consist of some topological manipulations (creating a new circuit, graph etc., and its transformation) and/or analytical manipulations of mathematical equations. In the case of complex topologies like matrix-reactance frequency converters, the choice of analytical method is very important. In this book the analytical approach is chosen. The circuit of Matrix-reactance frequency converters is modelled using average state-space methods. It should be noted that the models obtained as a result of averaging are continuously non-stationary ones. In order to obtain a stationary averaged state-space model, a two frequency dq transformation is used. A detailed analysis is presented in Sects. 4.2 and 4.3.

P. Szcześniak, *Three-Phase AC–AC Power Converters Based on Matrix Converter Topology*, Power Systems, DOI: 10.1007/978-1-4471-4896-8_4,

4.2 Averaged State-Space Model

The average state-space method was established by Middlebrook and Ćuk [27], and has been widely used for modelling DC–DC converters. However, as is demonstrated by publications [2–17, 19, 21–23, 30–33], this method can be applied to other types of power converters: AC–DC, AC–DC–AC. This modelling approach is based on the fact that the time-piecewise state equations are averaged over a switching cycle period to give a time continuous description. A theoretical foundation which provides a rigorous mathematical justification for widely used averaging methods in PWM converters is presented in [24, 27]. When using this approach, it is essential to have quite accurate models of the modulation process which is being considered.

An average state-space model is obtained with the following assumptions: all the switches are ideal (the voltage drop across the diode when forward biased is zero and no commutation losses in the transistor nor in the diode) and inductors and capacitors are linear. Furthermore, in the case of three phase converters an additional assumption should be taken into account: converter and sources are symmetrical and balanced.

The general form of the Averaged state-space equations is described by following set of equations [27]:

$$\frac{d\overline{\mathbf{x}}}{dt} = \mathbf{A}(d)\overline{\mathbf{x}} + \mathbf{B}(d), \tag{4.1}$$

$$\overline{\mathbf{y}} = \mathbf{C}(d)\overline{\mathbf{x}}, \tag{4.2}$$

where: $\overline{\mathbf{x}}$, $\overline{\mathbf{y}}$ are the vectors of the averaged state and output variables respectively; $\mathbf{A}(d)$, $\mathbf{B}(d)$, $\mathbf{C}(d)$, are the averaged state matrix, averaged input matrix and the averaged output matrix, respectively.

The state-space averaging method formulates the dynamic equations in state-space form for each of the switch configurations of the converter. The averaged model is then obtained by taking a weighted average of the system matrices, where the weighting factor for each switch configuration mode is its duty ratio defined as follows:

$$d_i = \frac{t_i}{T_{Seq}}, \tag{4.3}$$

where t_i-time period for ith switch configuration. From a mathematical point of view, the mathematical model of a power converter is represented by the sum of component for each switch configuration. Therefore, such systems can be mathematically described by a set of continuously differential equations.

The average state-space method applied to a matrix-reactance frequency converter is illustrated in Fig. 4.1 by block diagram. The inputs for the modelling algorithm are all subcircuits for allowed switch state combinations. In all topologies of matrix-reactance frequency converters, 28 switch states can be used. Then, there are defined differential equations for each of 28 switch configurations:

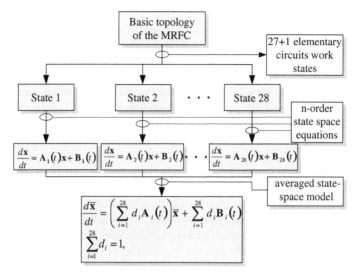

Fig. 4.1 Diagrammatic representation of the state-space averaging method for MRFCs

$$\frac{dx}{dt} = \mathbf{A}_i(t)\mathbf{x} + \mathbf{B}_i(t), \tag{4.4}$$

$$\mathbf{y} = \mathbf{C}_i(t)\mathbf{x}, \tag{4.5}$$

where: $\mathbf{x}$, $\mathbf{y}$ are the vectors of the state and output variables respectively; $A_i(t)$, $B_i(t)$, $C_i(t)$, are the state matrix, input matrix and the output matrix for ith switch configuration, respectively. The average state-space equations for matrix-reactance frequency converters can be represented by the following set of equations:

$$\frac{d\overline{\mathbf{x}}}{dt} = \mathbf{A}(d, t)\overline{\mathbf{x}} + \mathbf{B}(d, t), \tag{4.6}$$

$$\overline{\mathbf{y}} = \mathbf{C}(d, t)\overline{\mathbf{x}}, \tag{4.7}$$

where:

$$\sum_{i=1}^{28} d_i = 1, \tag{4.8}$$

$$\mathbf{A}(d, t) = \sum_{i=1}^{28} d_i \mathbf{A}_i(t), \tag{4.9}$$

$$\mathbf{B}(d, t) = \sum_{i=1}^{28} d_i \mathbf{B}_i(t), \tag{4.10}$$

$$\mathbf{C}(d, t) = \sum_{i=1}^{28} d_i \mathbf{C}_i(t), \tag{4.11}$$

The weight coefficient d_i is the degree of occurrence of all the possible configurations, and depends on the switch control strategy. Not all 28 switch configurations occur in each switch sequence period T_{Seq}. Equations (4.6)–(4.11) define the general form of the mathematical average state-space model for matrix-reactance frequency converters. It should be noted that defining modells from Eqs. (4.6)–(4.11) is very complex.

For Venturini control strategies, which are presented in this book for the control of MRFCs, matrices (4.9)–(4.11) are simplified. Considering that the power switches work with a high switching frequency, a low-frequency output voltage can be generated by modulating the duty ratio d_{jK} of the switches using their respective switching functions, as is presented in Chap. 2.3 and defined by the Eqs. (2.8)–(2.10) and shown in Fig. 2.42. Then the average duty factors of switch states in Eqs. (4.6)–(4.11) correspond to the Venturini duty factors d_{jK} defined by relations (3.2) and (3.3). Detailed average state-space equations are presented in Sect. 4.5. As mentioned above, the obtained model is continuously non-stationary, because coefficients d_{jK} are varied over time. In order to obtain a stationary Averaged state-space model, a two frequency dq transformation is used, as is presented in Sect. 4.3.

State-space averaging has been demonstrated to be an effective method for the analysis of MRFCs witch different control strategies. The popularity of the average state-space method is due largely to its clear and rational derivation, simple methodology, and demonstrated practical utility. The available modelling results provide tools for easier design and control of MRFCs.

4.3 Stationary State-Space Averaged Model: *dq* Transformation

Averaging and linearisation are typically distinguished as two separate steps. In order to obtain a stationary averaged state-space model (time-invariant), it is useful to introduce a dq transformation in a two frequency form (4.12)–(4.14) [5, 21, 22, 28–30]. It should be noted that the input and output system models have to be developed in their respective dq frames, which are defined for different pulsation ω and ω_L. This transformation is summarised by the diagram in Fig. 4.2.

$$\mathbf{K} = \begin{bmatrix} \mathbf{K}_S & \cdots & 0 \\ \vdots & \ddots & \vdots \\ 0 & \cdots & \mathbf{K}_L \end{bmatrix}, \tag{4.12}$$

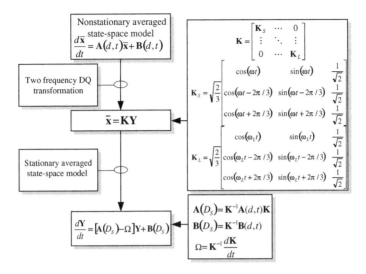

Fig. 4.2 Diagram representation of the two frequency *dq* transformation method

where:

$$\mathbf{K}_S = \sqrt{\frac{2}{3}} \begin{bmatrix} \cos(\omega t) & \sin(\omega t) & \frac{1}{\sqrt{2}} \\ \cos(\omega t + \frac{2\pi}{3}) & \sin(\omega t + \frac{2\pi}{3}) & \frac{1}{\sqrt{2}} \\ \cos(\omega t - \frac{2\pi}{3}) & \sin(\omega t - \frac{2\pi}{3}) & \frac{1}{\sqrt{2}} \end{bmatrix}, \tag{4.13}$$

$$\mathbf{K}_L = \sqrt{\frac{2}{3}} \begin{bmatrix} \cos(\omega_L t) & \sin(\omega_L t) & \frac{1}{\sqrt{2}} \\ \cos(\omega_L t + \frac{2\pi}{3}) & \sin(\omega_L t + \frac{2\pi}{3}) & \frac{1}{\sqrt{2}} \\ \cos(\omega_L t - \frac{2\pi}{3}) & \sin(\omega_L t - \frac{2\pi}{3}) & \frac{1}{\sqrt{2}} \end{bmatrix}, \tag{4.14}$$

where: $\mathbf{K}_S$, $\mathbf{K}_L$—are the *dq* transformation matrices defined for pulsation of the supply and load voltages, ω and ω_L respectively.

Furthermore, assuming that the converter circuit is symmetrical: $R_{LF1} = R_{LF2} = R_{LF3} = R_{FS}$, $L_{F1} = L_{F2} = L_{F3} = L_F$, $R_{LS1} = R_{LS2} = R_{LS3} = R_{LS}$, $L_{S1} = L_{S2} = L_{S3} = L_S$, $R_{LL1} = R_{LS2} = R_{LL3} = R_{LL}$, $L_{L1} = L_{L2} = L_{L3} = L_L$, $C_{F1} = C_{F2} = C_{F3} = C_F$, $C_{S1} = C_{S2} = C_{S3} = C_S$, $C_{L1} = C_{L2} = C_{L3} = C_L$, $R_{L1} = R_{L2} = R_{L3} = R_L$ and taking into consideration substitution (4.15) to (4.6) we obtain Eq. (4.16) with new state variables [5, 21, 22, 30].

$$\bar{\mathbf{x}} = \mathbf{K}\bar{\mathbf{x}}_{dq}, \tag{4.15}$$

$$\frac{d\mathbf{K}}{dt} + \mathbf{K}\frac{d\bar{\mathbf{x}}_{dq}}{dt} = \mathbf{A}(d, t)\mathbf{K}\bar{\mathbf{x}}_{dq} + \mathbf{B}(d, t). \tag{4.16}$$

Multiplying Eq. (4.16) by inverse matrix $\mathbf{K}^{-1}$ and taking into account (4.17) we obtain a general form of a time-invariant mathematical model based on the average state-space method for the matrix-reactance frequency converters in the dq frame expressed by (4.18).

$$\Omega = \mathbf{K}^{-1}\frac{d\mathbf{K}}{dt} = \begin{bmatrix} \Omega_S & \cdots & 0 \\ \vdots & \ddots & \vdots \\ 0 & \cdots & \Omega_L \end{bmatrix}, \tag{4.17}$$

$$\frac{d\overline{\mathbf{x}}_{dq}}{dt} = (\mathbf{A} - \Omega) + \mathbf{B}, \tag{4.18}$$

where:

$$\Omega_S = \begin{bmatrix} 0 & \omega & 0 \\ -\omega & 0 & 0 \\ 0 & 0 & 0 \end{bmatrix}, \qquad \Omega_L = \begin{bmatrix} 0 & \omega_L & 0 \\ -\omega_L & 0 & 0 \\ 0 & 0 & 0 \end{bmatrix}. \tag{4.19}$$

Defining a new matrix and vector of the analysed MRFCs parameters as in (4.20), finally we obtain a stationary averaged state-space model expressed by (4.21). The discussed method is presented, in details, in works [5, 6, 21, 22, 30].

$$\mathbf{A} = \mathbf{K}^{-1}\mathbf{A}(d, t)\mathbf{K}, \qquad \mathbf{B} = \mathbf{K}^{-1}\mathbf{B}(d, t). \tag{4.20}$$

$$\frac{d\overline{\mathbf{x}}_{dq}}{dt} = (\mathbf{A} - \Omega) + \mathbf{B}, \tag{4.21}$$

4.4 Solution of Stationary State-Space Averaged Equations

Equations (4.15)–(4.21) give a general description of the stationary state-space average model including three phase matrix-reactance frequency converters. The solution of the Eq. (4.21) as the value of vector $\overline{\mathbf{x}}_{dq}$ is described by (4.22).

$$\overline{\mathbf{x}}_{dq} = e^{(\mathbf{A}-\Omega)t}\overline{\mathbf{x}}_{dq}(0) + (\mathbf{A} - \Omega)^{-1}\big[e^{(\mathbf{A}-\Omega)t} - \mathbf{I}\big]\mathbf{B}, \tag{4.22}$$

where: $\mathbf{I}$ is the unit matrix, $\overline{\mathbf{x}}_{dq}(0)$—initial values of the transformed state variables. After rearranging (4.22), according to (4.15), we obtain a final description of the state variables in the MRFCs in the abc frame, which is expressed by (4.23) [5, 6, 21, 22, 30].

$$\overline{\mathbf{x}} = \mathbf{K}e^{(\mathbf{A}-\Omega)t}\overline{\mathbf{x}}_{dq}(0) + \mathbf{K}(\mathbf{A} - \Omega)^{-1}\big[e^{(\mathbf{A}-\Omega)t} - \mathbf{I}\big]\mathbf{B}. \tag{4.23}$$

The steady-state values of the averaged state variables obtained from (4.23) are described by (4.24)

$$\overline{\mathbf{x}}_{ust} = -\mathbf{K}(\mathbf{A} - \Omega)^{-1}\mathbf{B}. \tag{4.24}$$

The operation of the MRFC topologies given in Figs. 3.6, 3.7, 3.8, 3.9 and 3.10 can be analyzed with the help of average differential Eqs. (4.24) and (4.7). This solution can be used for any control strategy of MRFCs.

4.5 Mathematical Models of Matrix Reactance Frequency Converters

As explained above, the mathematical average state-space models are expressed by Eqs. (4.6) and (4.7). The stationary model of a MRFC after dq transformation is given by Eq. (4.21). The output equations are in the same form and is defined by (4.7).

To illustrate the general principles and mathematical models so far described, consider the simple MRFC-I-b-b circuit shown in Fig. 4.3. In this figure all the considered voltages and currents are indicated. It is assumed that the converter circuit is symmetrical and the power supply is balanced. The vector of the average output values $\overline{\mathbf{y}}$ is arbitrarily defined. Let us consider output variables described by voltages between source star point and star point of the capacitors and inductors $\overline{\mathbf{y}} = [u_{N1}, u_{N2}, u_{N3}, u_{N4}]^T$ as shown in Fig. 4.3. The source star point is grounded and is used as a reference potential. In this three phase three wire system (Fig. 4.3), voltages $u_{N1}, u_{N2}, u_{N3}, u_{N4}$, can be expressed from the other voltages.

Assuming that:

$$\begin{aligned}
u_{F1} + u_{F2} + u_{F3} &= 0, \\
u_{D1} + u_{D2} + u_{D3} &= 0, \\
u_{C1} + u_{C2} + u_{C3} &= 0, \\
u_{R1} + u_{R2} + u_{R3} &= 0,
\end{aligned} \tag{4.25}$$

and

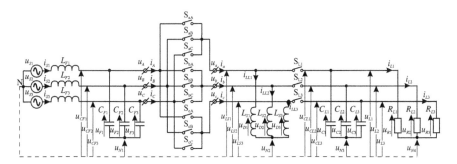

Fig. 4.3 Voltage and current description for matrix-reactance frequency converter based on buck-boost I topology

$$u_{CF1} = u_{F1} + u_{N1} = 0, \ u_{CF2} = u_{F2} + u_{N1} = 0, \ u_{CF3} = u_{F3} + u_{N1} = 0,$$
$$u_{LS1} = u_{D1} + u_{N2} = 0, \ u_{LS2} = u_{D2} + u_{N2} = 0, \ u_{LS3} = u_{D3} + u_{N2} = 0,$$
$$u_{CL1} = u_{C1} + u_{N3} = 0, \ u_{CL2} = u_{C2} + u_{N3} = 0, \ u_{CL3} = u_{C3} + u_{N3} = 0,$$
$$u_{L1} = u_{R1} + u_{N4} = 0, \quad u_{L2} = u_{R2} + u_{N4} = 0, \quad u_{L3} = u_{R3} + u_{N4} = 0,$$
$$(4.26)$$

after substituting Eqs. (4.26) into (4.25) becomes:

$$u_{N1} = (u_{CF1} + u_{CF2} + u_{CF3})/3,$$
$$u_{N2} = (u_{LS1} + u_{LS2} + u_{LS3})/3,$$
$$u_{N3} = (u_{CL1} + u_{CL2} + u_{CL3})/3,$$
$$u_{N4} = (u_{L1} + u_{L2} + u_{L3})/3.$$
$$(4.27)$$

Equations (4.25)–(4.27) can, in the usual manner, be generalised for other MRFC topologies [30].

The averaged state-space equations in detailed form for the discussed MRFC (Fig. 4.3) are obtained by introducing the modified Venturini control strategy described by (3.1) and (3.2) and Fig. 3.17a, to Eqs. (4.6) and (4.7). Continuously non-stationary equations may be described by the set Eqs. (4.28). These equations are derived from the Kirchhoff Current Law and the Kirchhoff Voltage Law. Matrix $A(d, t)$ has m rows and columns, and its dimensions depend on the number of converter reactive elements (number of inductances and capacitances). It defines the relationship between the state variable signals and specifies the operation of the power switches (duty factors) in the frequency converter. These duty factors correspond to the actual modulating signals sent to the control modulators. For the presented model, duty factors are generated using the modified Venturini algorithm presented in [30]. Matrix $B(d, t)$ defines all the voltage and current sources in the modelling system. Furthermore, the average output values are described by expression (4.29) [30]. Matrix $C(d, t)$ defines the relation between state variables and an arbitrarily defined vector of output variables. The presented model is determined for systems of MRPC working in an open loop feedback control system, and taking into consideration only simplify models of converter's reactive elements. The inductors are modelled as inductances with series connected resistance. Capacitors, power switches, voltage sources and load are ideal, without parasitic capacitive, inductances and resistances.

Thus, this averaged description can be used for any control approach if the switching functions can also be obtained. An example of modelling an MRFC-I-b-b SVM is presented in paper [34].

$$
\begin{bmatrix}
\dfrac{d\bar{i}_{S1}}{dt} \\[4pt]
\dfrac{d\bar{i}_{S2}}{dt} \\[4pt]
\dfrac{d\bar{i}_{S3}}{dt} \\[4pt]
\dfrac{d\bar{i}_{LS1}}{dt} \\[4pt]
\dfrac{d\bar{i}_{LS2}}{dt} \\[4pt]
\dfrac{d\bar{i}_{LS3}}{dt} \\[4pt]
\dfrac{d\bar{u}_{CF1}}{dt} \\[4pt]
\dfrac{d\bar{u}_{CF2}}{dt} \\[4pt]
\dfrac{d\bar{u}_{CF3}}{dt} \\[4pt]
\dfrac{d\bar{u}_{L1}}{dt} \\[4pt]
\dfrac{d\bar{u}_{L2}}{dt} \\[4pt]
\dfrac{d\bar{u}_{L3}}{dt}
\end{bmatrix}
=
\begin{bmatrix}
\frac{-R_{LF}}{L_F} & 0 & 0 & 0 & 0 & 0 & \frac{-1}{L_F} & 0 & 0 & 0 & 0 & 0 \\[4pt]
0 & \frac{-R_{LF}}{L_F} & 0 & 0 & 0 & 0 & 0 & \frac{-1}{L_F} & 0 & 0 & 0 & 0 \\[4pt]
0 & 0 & \frac{-R_{LF}}{L_F} & 0 & 0 & 0 & 0 & 0 & \frac{-1}{L_F} & 0 & 0 & 0 \\[4pt]
0 & 0 & 0 & \frac{-R_{LS}}{L_S} & 0 & 0 & \frac{d_{aA}}{L_S} & \frac{d_{aB}}{L_S} & \frac{d_{aC}}{L_S} & \frac{1-D_S}{L_S} & 0 & 0 \\[4pt]
0 & 0 & 0 & 0 & \frac{-R_{LS}}{L_S} & 0 & \frac{d_{bA}}{L_S} & \frac{d_{bB}}{L_S} & \frac{d_{bC}}{L_S} & 0 & \frac{1-D_S}{L_S} & 0 \\[4pt]
0 & 0 & 0 & 0 & 0 & \frac{-R_{LS}}{L_S} & \frac{d_{cA}}{L_S} & \frac{d_{cB}}{L_S} & \frac{d_{cC}}{L_S} & 0 & 0 & \frac{1-D_S}{L_S} \\[4pt]
\frac{1}{C_F} & 0 & 0 & \frac{-d_{aA}}{C_F} & \frac{-d_{aB}}{C_F} & \frac{-d_{aC}}{C_F} & 0 & 0 & 0 & 0 & 0 & 0 \\[4pt]
0 & \frac{1}{C_F} & 0 & \frac{-d_{bA}}{C_F} & \frac{-d_{bB}}{C_F} & \frac{-d_{bC}}{C_F} & 0 & 0 & 0 & 0 & 0 & 0 \\[4pt]
0 & 0 & \frac{1}{C_F} & \frac{-d_{cA}}{C_F} & \frac{-d_{cB}}{C_F} & \frac{-d_{cC}}{C_F} & 0 & 0 & 0 & 0 & 0 & 0 \\[4pt]
0 & 0 & 0 & \frac{D_S-1}{C_L} & 0 & 0 & 0 & 0 & 0 & \frac{-1}{R_L C_L} & 0 & 0 \\[4pt]
0 & 0 & 0 & 0 & \frac{D_S-1}{C_L} & 0 & 0 & 0 & 0 & 0 & \frac{-1}{R_L C_L} & 0 \\[4pt]
0 & 0 & 0 & 0 & 0 & \frac{D_S-1}{C_L} & 0 & 0 & 0 & 0 & 0 & \frac{-1}{R_L C_L}
\end{bmatrix}
\begin{bmatrix}
\bar{i}_{S1} \\[4pt]
\bar{i}_{S2} \\[4pt]
\bar{i}_{S3} \\[4pt]
\bar{i}_{LS1} \\[4pt]
\bar{i}_{LS2} \\[4pt]
\bar{i}_{LS3} \\[4pt]
\bar{u}_{CF1} \\[4pt]
\bar{u}_{CF2} \\[4pt]
\bar{u}_{CF3} \\[4pt]
\bar{u}_{L1} \\[4pt]
\bar{u}_{L2} \\[4pt]
\bar{u}_{L3}
\end{bmatrix}
+
\begin{bmatrix}
\frac{u_{S1}}{L_F} \\[4pt]
\frac{u_{S2}}{L_F} \\[4pt]
\frac{u_{S3}}{L_F} \\[4pt]
0 \\[4pt]
0 \\[4pt]
0 \\[4pt]
0 \\[4pt]
0 \\[4pt]
0 \\[4pt]
0 \\[4pt]
0 \\[4pt]
0
\end{bmatrix}
\tag{4.28}
$$

$$
\begin{bmatrix} \bar{u}_{N1} \\ \bar{u}_{N2} \\ \bar{u}_{N3} \\ \bar{u}_{N4} \end{bmatrix} = \frac{1}{3} \begin{bmatrix} 0\,0\,0\,0\,0\,0 & 1 & 1 & 1 & 0 & 0 & 0 \\ 0\,0\,0\,0\,0\,0 & D_S & D_S & D_S & (1-D_S) & (1-D_S) & (1-D_S) \\ 0\,0\,0\,0\,0\,0 & 0 & 0 & 0 & 1 & 1 & 1 \\ 0\,0\,0\,0\,0\,0 & 0 & 0 & 0 & 1 & 1 & 1 \end{bmatrix} \bar{x} \quad (4.29)
$$

Knowing the continuously non-stationary equations, it is possible to obtain continuously stationary equations with the use of the algorithm shown in Fig. 4.2 and expressed by Eqs. (4.15)–(4.20). Equation (4.28) may be written by (4.32), where the dq transformation matrix is expressed by (4.31) [30]. The construction of the matrix $\mathbf{K}$ depends on the frequency of the state variables. If the frequency of capacitor voltages or inductor currents is equal to the frequency of source voltages then appropriate elements of matrix $\mathbf{K}$ will by defined for pulsation ω_S. Otherwise, they will be defined for output pulsation ω_L. Sub-matrices $\mathbf{K}_S$, and $\mathbf{K}_L$ are presented by the expressions (4.13) and (4.14) respectively.

$$
\mathbf{K} = \begin{bmatrix} \mathbf{K}_S & 0 & 0 & 0 \\ 0 & \mathbf{K}_L & 0 & 0 \\ 0 & 0 & \mathbf{K}_S & 0 \\ 0 & 0 & 0 & \mathbf{K}_L \end{bmatrix} \quad (4.30)
$$

$$
\frac{d}{dt}
\begin{bmatrix}
\bar{i}_{S1dq} \\
\bar{i}_{S2dq} \\
\bar{i}_{S3dq} \\
\bar{i}_{LS1dq} \\
\bar{i}_{LS2dq} \\
\bar{i}_{LS3dq} \\
\bar{u}_{CF1dq} \\
\bar{u}_{CF2dq} \\
\bar{u}_{CF3dq} \\
\bar{u}_{L1dq} \\
\bar{u}_{L2dq} \\
\bar{u}_{L3dq}
\end{bmatrix}
=
\mathbf{A}
\begin{bmatrix}
\bar{i}_{S1dq} \\
\bar{i}_{S2dq} \\
\bar{i}_{S3dq} \\
\bar{i}_{LS1dq} \\
\bar{i}_{LS2dq} \\
\bar{i}_{LS3dq} \\
\bar{u}_{CF1dq} \\
\bar{u}_{CF2dq} \\
\bar{u}_{CF3dq} \\
\bar{u}_{L1dq} \\
\bar{u}_{L2dq} \\
\bar{u}_{L3dq}
\end{bmatrix}
+
\begin{bmatrix}
\sqrt{\tfrac{3}{2}}\,\dfrac{U_m}{L_F} \\
0 \\
0 \\
0 \\
0 \\
0 \\
0 \\
0 \\
0 \\
0 \\
0 \\
0
\end{bmatrix}
\tag{4.31}
$$

where

$$
\mathbf{A}=
\begin{bmatrix}
-\dfrac{R_{LF}}{L_F} & -\omega & 0 & 0 & 0 & 0 & -\dfrac{1}{L_F} & 0 & 0 & 0 & 0 & 0\\[4pt]
\omega & -\dfrac{R_{LF}}{L_F} & 0 & 0 & 0 & 0 & 0 & -\dfrac{1}{L_F} & 0 & 0 & 0 & 0\\[4pt]
0 & 0 & -\dfrac{R_{LF}}{L_F} & 0 & 0 & 0 & 0 & 0 & -\dfrac{1}{L_F} & 0 & 0 & 0\\[4pt]
0 & 0 & 0 & -\dfrac{R_{LS}}{L_S} & -\omega_L & 0 & \dfrac{1-D_S}{L_S} & 0 & 0 & -\dfrac{D_S}{L_S} & 0 & 0\\[4pt]
0 & 0 & 0 & \omega_L & -\dfrac{R_{LS}}{L_S} & 0 & 0 & \dfrac{1-D_S}{L_S} & 0 & 0 & -\dfrac{qD_S}{L_S} & 0\\[4pt]
0 & 0 & 0 & 0 & 0 & -\dfrac{R_{LS}}{L_S} & 0 & 0 & \dfrac{1-D_S}{L_S} & 0 & 0 & \dfrac{qD_S}{L_S}\\[4pt]
\dfrac{1}{C_F} & 0 & 0 & -\dfrac{D_S}{C_F} & 0 & 0 & 0 & -\omega & 0 & 0 & 0 & 0\\[4pt]
0 & \dfrac{1}{C_F} & 0 & 0 & -\dfrac{qD_S}{C_F} & 0 & \omega & 0 & 0 & 0 & 0 & 0\\[4pt]
0 & 0 & \dfrac{1}{C_F} & 0 & 0 & -\dfrac{qD_S}{C_F} & 0 & 0 & 0 & 0 & 0 & 0\\[4pt]
0 & 0 & 0 & \dfrac{D_S-1}{C_L} & 0 & 0 & 0 & 0 & 0 & -\dfrac{1}{R_L C_L} & -\omega_L & 0\\[4pt]
0 & 0 & 0 & 0 & \dfrac{D_S-1}{C_L} & 0 & 0 & 0 & 0 & \omega_L & -\dfrac{1}{R_L C_L} & 0\\[4pt]
0 & 0 & 0 & 0 & 0 & \dfrac{D_S-1}{C_L} & 0 & 0 & 0 & 0 & 0 & -\dfrac{1}{R_L C_L}
\end{bmatrix}
$$

Substituting (4.31) into (4.29), one can define the average output variables (4.32). After rearrangement there are obtained the voltages u_{N1}, u_{N2}, u_{N3}, u_{N4} equal zero. This means that in a MRFC with modified Venturini control strategy (3.1), (3.2), the input/output waveforms are sinusoidal. In an improved Venturini and SVM modulation these relationships are different. The averaged voltages u_{N1}, u_{N2}, u_{N3}, u_{N4}, do not equal zero [31, 34]. In all of the MRFC topologies with modified Venturini modulation, the averaged voltages between source star point and star point of the capacitors and inductors are equal to zero as shown in the work [30]. The average output variables model for other topologies will not be shown.

$$\begin{bmatrix} \bar{u}_{N1} \\ \bar{u}_{N2} \\ \bar{u}_{N3} \\ \bar{u}_{N4} \end{bmatrix} = \frac{1}{3} \begin{bmatrix} \bar{u}_{CF1} + \bar{u}_{CF2} + \bar{u}_{CF3} \\ D_S(\bar{u}_{CF1} + \bar{u}_{CF2} + \bar{u}_{CF3}) + (1 - D_S)(\bar{u}_{L1} + \bar{u}_{L2} + \bar{u}_{L3}) \\ \bar{u}_{L1} + \bar{u}_{L2} + \bar{u}_{L3} \\ \bar{u}_{L1} + \bar{u}_{L2} + \bar{u}_{L3} \end{bmatrix} = \begin{bmatrix} 0 \\ 0 \\ 0 \\ 0 \end{bmatrix}$$

(4.32)

The concise form of non-stationary and stationary average state-space models described by the Eqs. (4.6) and (4.21) respectively and transformation matrices **K** for all MRFC topologies (Figs. 3.6, 3.7, 3.8, 3.9 and 3.10) are collected in Tables 4.1, 4.2, 4.3, 4.4, 4.5, 4.6, 4.7 and 4.8. All submatrices are collated in Table 4.9 at the end of this chapter. As is presented in Sect. 3.5, four switching patterns are given (Figs. 3.17 and 3.18) for the family of MRFCs with modified Venturini control strategy [30]. Two low frequency modulation matrices are used in the control of switches (3.1)–(3.3). Both modulation matrix and switching pattern are taken into consideration in the derivation process of the mathematical averaged state-space model. For topologies of MRFC with voltage source matrix converter (MRFC-I-b-b, MRFC-II-c, MRFC-I-z, MRFC-II-z), averaged duty factors of switch state functions are defined by the expressions (3.2), whereas for a MRFC with current source matrix converters (MRFC-II-b-b, MRFC-I-c, MRFC-I-s, MRFC-II-s, MRFC-b) there are used expressions defined by the Eq. (3.3) [5, 30].

The determination of mathematical averaged state-space models of MRFCs in the presented way is a simple task, which requires only a mathematical transformation without any circuit transformation.

As shown above, the solution of stationary averaged state models is specified by Eqs. (4.23) and (4.24) for transient and steady state respectively. An exemplary solution, showing input current and output voltage symbolic expressions for MRFC-I-b-b topology, can be expressed as (4.33)–(4.35) [30]. The relations (4.33)–(4.35) is complicated, and its analysis is difficult. However, the dynamic development of computer systems and mathematical software contribute to the formation of many advanced tools which bring new perspectives to the solution of mathematical problems in the symbolic and numerical approach. The presented Averaged state-space modelling approach is relatively easy to use in formulating the basic averaged set equations, requires only a small number of mathematical transformations and circuit transformation is not required. In the next chapter the analysis of MRFC properties in steady and transient states based on theoretical analysis will be presented.

Table 4.1 Mathematical Averaged state-space model of MRPC-II-b-b

Non-stationary average state-space set equations

$$
\begin{bmatrix}
\frac{d\bar{i}_S}{dt} \\[4pt]
\frac{d\bar{i}_{LS}}{dt} \\[4pt]
\frac{d\bar{u}_{CF}}{dt} \\[4pt]
\frac{d\bar{u}_L}{dt}
\end{bmatrix}
=
\begin{bmatrix}
-\mathbf{L}_F\mathbf{R}_{LF} & 0 & -\mathbf{L}_F & 0 \\
0 & -\mathbf{L}_S\mathbf{R}_{LS} & \mathbf{L}_S\mathbf{D}_1 & \mathbf{L}_S\mathbf{D}_M \\
\mathbf{C}_F & -\mathbf{C}_F\mathbf{D}_1 & 0 & 0 \\
0 & -\mathbf{C}_L\mathbf{D}_M^T & 0 & -\mathbf{C}_L\mathbf{R}_L
\end{bmatrix}
\begin{bmatrix}
\bar{i}_S \\[4pt]
\bar{i}_{LS} \\[4pt]
\bar{u}_{CF} \\[4pt]
\bar{u}_L
\end{bmatrix}
+
\begin{bmatrix}
\mathbf{L}_F\mathbf{u}_S \\[4pt]
0 \\[4pt]
0 \\[4pt]
0
\end{bmatrix}
$$

Two frequency dq transformation matrix

$$
\mathbf{K} =
\begin{bmatrix}
\mathbf{K}_S & 0 & 0 & 0 \\
0 & \mathbf{K}_S & 0 & 0 \\
0 & 0 & \mathbf{K}_S & 0 \\
0 & 0 & 0 & \mathbf{K}_L
\end{bmatrix}
$$

Stationary average state-space set equations

$$
\begin{bmatrix}
\frac{d\bar{i}_{Sdq}}{dt} \\[4pt]
\frac{d\bar{i}_{LSdq}}{dt} \\[4pt]
\frac{d\bar{u}_{CFdq}}{dt} \\[4pt]
\frac{d\bar{u}_{Ldq}}{dt}
\end{bmatrix}
=
\begin{bmatrix}
-\mathbf{L}_F\mathbf{R}_{LF} - \Omega_S & 0 & -\mathbf{L}_F & 0 \\
0 & -\mathbf{L}_S\mathbf{R}_{LS} - \Omega_S & \mathbf{L}_S\mathbf{D}_1 & \mathbf{L}_S\mathbf{D}_{M1dq} \\
\mathbf{C}_F & -\mathbf{C}_F\mathbf{D}_1 & -\Omega_S & 0 \\
0 & -\mathbf{C}_L\mathbf{D}_{M1dq}^T & 0 & -\mathbf{C}_L\mathbf{R}_L - \Omega_L
\end{bmatrix}
\begin{bmatrix}
\bar{i}_{Sdq} \\[4pt]
\bar{i}_{LSdq} \\[4pt]
\bar{u}_{CFdq} \\[4pt]
\bar{u}_{Ldq}
\end{bmatrix}
+
\begin{bmatrix}
\mathbf{L}_F\mathbf{u}_{Sdq} \\[4pt]
0 \\[4pt]
0 \\[4pt]
0
\end{bmatrix}
$$

Table 4.2 Mathematical Averaged state-space model of MRPC-I-c

Non-stationary average state-space set equations

$$
\begin{bmatrix}
\frac{d\bar{i}_S}{dt} \\[4pt]
\frac{d\bar{i}_{LL}}{dt} \\[4pt]
\frac{d\bar{u}_{CS}}{dt} \\[4pt]
\frac{d\bar{u}_L}{dt}
\end{bmatrix}
=
\begin{bmatrix}
-\mathbf{L}_S\mathbf{R}_{LS} & 0 & -\mathbf{L}_S\mathbf{D}_M & 0 \\
0 & -\mathbf{L}_L\mathbf{R}_{LL} & \mathbf{L}_L\mathbf{D}_1 & -\mathbf{L}_L \\
\mathbf{C}_S\mathbf{D}_M^T & \mathbf{C}_S\mathbf{D}_1 & 0 & 0 \\
0 & \mathbf{C}_L & 0 & -\mathbf{C}_L\mathbf{R}_L
\end{bmatrix}
\begin{bmatrix}
\bar{i}_S \\[4pt]
\bar{i}_{LL} \\[4pt]
\bar{u}_{CS} \\[4pt]
\bar{u}_L
\end{bmatrix}
+
\begin{bmatrix}
\mathbf{L}_S\mathbf{u}_S \\[4pt]
0 \\[4pt]
0 \\[4pt]
0
\end{bmatrix}
$$

Two frequency dq transformation matrix

$$
\mathbf{K} =
\begin{bmatrix}
\mathbf{K}_S & 0 & 0 & 0 \\
0 & \mathbf{K}_L & 0 & 0 \\
0 & 0 & \mathbf{K}_L & 0 \\
0 & 0 & 0 & \mathbf{K}_L
\end{bmatrix}
$$

Stationary average state-space set equations

$$
\begin{bmatrix}
\frac{d\bar{i}_{Sdq}}{dt} \\[4pt]
\frac{d\bar{i}_{LLdq}}{dt} \\[4pt]
\frac{d\bar{u}_{CSdq}}{dt} \\[4pt]
\frac{d\bar{u}_{Ldq}}{dt}
\end{bmatrix}
=
\begin{bmatrix}
-\mathbf{L}_S\mathbf{R}_{LS} - \Omega_S & 0 & -\mathbf{L}_S\mathbf{D}_{Mdq1} & 0 \\
0 & -\mathbf{L}_L\mathbf{R}_{LL} - \Omega_L & \mathbf{L}_L\mathbf{D}_1 & -\mathbf{L}_L \\
\mathbf{C}_S\mathbf{D}_{Mdq1}^T & \mathbf{C}_S\mathbf{D}_1 & -\Omega_L & 0 \\
0 & \mathbf{C}_L & 0 & -\mathbf{C}_L\mathbf{R}_L - \Omega_L
\end{bmatrix}
\begin{bmatrix}
\bar{i}_{Sdq} \\[4pt]
\bar{i}_{LLdq} \\[4pt]
\bar{u}_{CSdq} \\[4pt]
\bar{u}_{Ldq}
\end{bmatrix}
+
\begin{bmatrix}
\mathbf{L}_S\mathbf{u}_{Sdq} \\[4pt]
0 \\[4pt]
0 \\[4pt]
0
\end{bmatrix}
$$

Table 4.3 Mathematical Averaged state-space model of MRPC-II-c

Non-stationary average state-space set equations

$$
\begin{bmatrix} \frac{d\bar{i}_S}{dt} \\ \frac{d\bar{i}_{LL}}{dt} \\ \frac{d\bar{u}_{CS}}{dt} \\ \frac{d\bar{u}_L}{dt} \end{bmatrix} =
\begin{bmatrix}
-L_S R_{LS} & 0 & -L_S D_2 & 0 \\
0 & -L_L R_{LL} & -L_L D_M & -L_L \\
C_S D_2 & C_S D_M^T & 0 & 0 \\
0 & C_L & 0 & -C_L R_L
\end{bmatrix}
\begin{bmatrix} \bar{i}_S \\ \bar{i}_{LL} \\ \bar{u}_{CS} \\ \bar{u}_L \end{bmatrix} +
\begin{bmatrix} L_S u_S \\ 0 \\ 0 \\ 0 \end{bmatrix}
$$

Two frequency dq transformation matrix

$$
K = \begin{bmatrix}
K_S & 0 & 0 & 0 \\
0 & K_L & 0 & 0 \\
0 & 0 & K_S & 0 \\
0 & 0 & 0 & K_L
\end{bmatrix}
$$

Stationary average state-space set equations

$$
\begin{bmatrix} \frac{d\bar{i}_{Sdq}}{dt} \\ \frac{d\bar{i}_{LLdq}}{dt} \\ \frac{d\bar{u}_{CSdq}}{dt} \\ \frac{d\bar{u}_{Ldq}}{dt} \end{bmatrix} =
\begin{bmatrix}
-L_S R_{LS} - \Omega_S & 0 & -L_S D_2 & 0 \\
0 & -L_L R_{LL} - \Omega_L & -L_L D_{Mdq} & -L_L \\
C_S D_2 & C_S D_{Mdq}^T & -\Omega_S & 0 \\
0 & C_L & 0 & -C_L R_L - \Omega_L
\end{bmatrix}
\begin{bmatrix} \bar{i}_{Sdq} \\ \bar{i}_{LLdq} \\ \bar{u}_{CSdq} \\ \bar{u}_{Ldq} \end{bmatrix} +
\begin{bmatrix} L_S u_{Sdq} \\ 0 \\ 0 \\ 0 \end{bmatrix}
$$

Table 4.4 Mathematical Averaged state-space model of MRPC-I-z

Non-stationary average state-space set equations

$$
\begin{bmatrix} \frac{d\bar{i}_S}{dt} \\ \frac{d\bar{i}_{LS}}{dt} \\ \frac{d\bar{i}_{LL}}{dt} \\ \frac{d\bar{u}_{CF}}{dt} \\ \frac{d\bar{u}_{CS}}{dt} \\ \frac{d\bar{u}_L}{dt} \end{bmatrix} =
\begin{bmatrix}
-L_F R_{LF} & 0 & 0 & -L_F & 0 & 0 \\
0 & -L_S R_{LS} & 0 & L_S D_M & L_S D_2 & 0 \\
0 & 0 & -L_L R_{LL} & L_L D_M & -L_L D_1 & -L_L \\
C_F & -C_F D_M^T & -C_F D_M^T & 0 & 0 & 0 \\
0 & -C_S D_2 & C_S D_1 & 0 & 0 & 0 \\
0 & 0 & C_L & 0 & 0 & -C_L R_L
\end{bmatrix}
\begin{bmatrix} \bar{i}_S \\ \bar{i}_{LS} \\ \bar{i}_{LL} \\ \bar{u}_{CF} \\ \bar{u}_{CS} \\ \bar{u}_L \end{bmatrix} +
\begin{bmatrix} L_F u_S \\ 0 \\ 0 \\ 0 \\ 0 \\ 0 \end{bmatrix}
$$

Two frequency dq transformation matrix

$$
K = \begin{bmatrix}
K_S & 0 & 0 & 0 & 0 & 0 \\
0 & K_L & 0 & 0 & 0 & 0 \\
0 & 0 & K_L & 0 & 0 & 0 \\
0 & 0 & 0 & K_S & 0 & 0 \\
0 & 0 & 0 & 0 & K_L & 0 \\
0 & 0 & 0 & 0 & 0 & K_L
\end{bmatrix}
$$

Stationary average state-space set equations

$$
\begin{bmatrix} \frac{d\bar{i}_{Sdq}}{dt} \\ \frac{d\bar{i}_{LSdq}}{dt} \\ \frac{d\bar{i}_{LLdq}}{dt} \\ \frac{d\bar{u}_{CFdq}}{dt} \\ \frac{d\bar{u}_{CSdq}}{dt} \\ \frac{d\bar{u}_{Ldq}}{dt} \end{bmatrix} =
\begin{bmatrix}
-L_F R_{LF} - \Omega_S & 0 & 0 & -L_F & 0 & 0 \\
0 & -L_S R_{LS} - \Omega_L & 0 & L_S D_{Mdq} & L_S D_2 & 0 \\
0 & 0 & -L_L R_{LL} - \Omega_L & L_L D_{Mdq} & -L_L D_1 & -L_L \\
C_F & -C_F D_{Mdq}^T & -C_F D_{Mdq}^T & -\Omega_S & 0 & 0 \\
0 & -C_S D_2 & C_S D_1 & 0 & -\Omega_L & 0 \\
0 & 0 & C_L & 0 & 0 & -C_L R_L - \Omega_L
\end{bmatrix}
\begin{bmatrix} \bar{i}_{Sdq} \\ \bar{i}_{LSdq} \\ \bar{i}_{LLdq} \\ \bar{u}_{CFdq} \\ \bar{u}_{CSdq} \\ \bar{u}_{Ldq} \end{bmatrix} +
\begin{bmatrix} L_F u_{Sdq} \\ 0 \\ 0 \\ 0 \\ 0 \\ 0 \end{bmatrix}
$$

Table 4.5 Mathematical Averaged state-space model of MRPC-II-z

Non-stationary average state-space set equations

$$
\begin{bmatrix} \frac{d\bar{i}_S}{dt} \\ \frac{d\bar{i}_{LS}}{dt} \\ \frac{d\bar{i}_{LL}}{dt} \\ \frac{d\bar{u}_{CF}}{dt} \\ \frac{d\bar{u}_{CS}}{dt} \\ \frac{d\bar{u}_L}{dt} \end{bmatrix} =
\begin{bmatrix}
-L_F R_{LF} & 0 & 0 & -L_F & 0 & 0 \\
0 & -L_S R_{LS} & 0 & L_S D_1 & L_S D_2 & 0 \\
0 & 0 & -L_L R_{LL} & L_L D_M & -L_L D_M & -L_L \\
C_F & -C_F D_1 & -C_F D_M^T & 0 & 0 & 0 \\
0 & -C_S D_2 & C_S D_M^T & 0 & 0 & 0 \\
0 & 0 & C_L & 0 & 0 & -C_L R_L
\end{bmatrix}
\begin{bmatrix} \bar{i}_S \\ \bar{i}_{LS} \\ \bar{i}_{LL} \\ \bar{u}_{CF} \\ \bar{u}_{CS} \\ \bar{u}_L \end{bmatrix} +
\begin{bmatrix} L_F u_S \\ 0 \\ 0 \\ 0 \\ 0 \\ 0 \end{bmatrix}
$$

Two frequency dq transformation matrix

$$
K =
\begin{bmatrix}
K_S & 0 & 0 & 0 & 0 & 0 \\
0 & K_S & 0 & 0 & 0 & 0 \\
0 & 0 & K_L & 0 & 0 & 0 \\
0 & 0 & 0 & K_S & 0 & 0 \\
0 & 0 & 0 & 0 & K_S & 0 \\
0 & 0 & 0 & 0 & 0 & K_L
\end{bmatrix}
$$

Stationary average state-space set equations

$$
\begin{bmatrix} \frac{d\bar{i}_{Sdq}}{dt} \\ \frac{d\bar{i}_{LSdq}}{dt} \\ \frac{d\bar{i}_{LLdq}}{dt} \\ \frac{d\bar{u}_{CFdq}}{dt} \\ \frac{d\bar{u}_{CSdq}}{dt} \\ \frac{d\bar{u}_{Ldq}}{dt} \end{bmatrix} =
\begin{bmatrix}
-L_F R_{LF}-\Omega_S & 0 & 0 & L_F & 0 & 0 \\
0 & -L_S R_{LS}-\Omega_S & 0 & L_S D_1 & L_S D_2 & 0 \\
0 & 0 & -L_L R_{LL}-\Omega_L & L_L D_{Mdq} & -L_L D_{Mdq} & -L_L \\
C_F & -C_F D_1 & -C_F D_{Mdq}^T & -\Omega_S & 0 & 0 \\
0 & -C_S D_2 & C_S D_{Mdq}^T & 0 & -\Omega_S & 0 \\
0 & 0 & C_L & 0 & 0 & -C_L R_L-\Omega_L
\end{bmatrix}
\begin{bmatrix} \bar{i}_{Sdq} \\ \bar{i}_{LSdq} \\ \bar{i}_{LLdq} \\ \bar{u}_{CFdq} \\ \bar{u}_{CSdq} \\ \bar{u}_{Ldq} \end{bmatrix} +
\begin{bmatrix} L_F u_{Sdq} \\ 0 \\ 0 \\ 0 \\ 0 \\ 0 \end{bmatrix}
$$

Table 4.6 Mathematical Averaged state-space model of MRPC-I-s

Non-stationary average state-space set equations

$$
\begin{bmatrix} \frac{d\bar{i}_S}{dt} \\ \frac{d\bar{i}_{LL}}{dt} \\ \frac{d\bar{u}_{CS}}{dt} \\ \frac{d\bar{u}_L}{dt} \end{bmatrix} =
\begin{bmatrix}
-L_S R_{LS} & 0 & -L_S D_M & -L_S D_M \\
0 & -L_L R_{LL} & -L_L D_1 & L_L D_2 \\
C_S D_M^T & C_S D_1 & 0 & 0 \\
C_L D_M^T & -C_L D_2 & 0 & -C_L R_L
\end{bmatrix}
\begin{bmatrix} \bar{i}_S \\ \bar{i}_{LL} \\ \bar{u}_{CS} \\ \bar{u}_L \end{bmatrix} +
\begin{bmatrix} L_S u_S \\ 0 \\ 0 \\ 0 \end{bmatrix}
$$

Two frequency dq transformation matrix

$$
K =
\begin{bmatrix}
K_S & 0 & 0 & 0 \\
0 & K_L & 0 & 0 \\
0 & 0 & K_L & 0 \\
0 & 0 & 0 & K_L
\end{bmatrix}
$$

Stationary average state-space set equations

$$
\begin{bmatrix} \frac{d\bar{i}_{Sdq}}{dt} \\ \frac{d\bar{i}_{LLdq}}{dt} \\ \frac{d\bar{u}_{CSdq}}{dt} \\ \frac{d\bar{u}_{Ldq}}{dt} \end{bmatrix} =
\begin{bmatrix}
-L_S R_{LS}-\Omega_S & 0 & -L_S D_{Mdq1} & -L_S D_{Mdq1} \\
0 & -L_L R_{LL}-\Omega_L & -L_L D_1 & L_L D_2 \\
C_S D_{Mdq1}^T & C_S D_1 & -\Omega_L & 0 \\
C_L D_{Mdq1}^T & -C_L D_2 & 0 & -C_L R_L-\Omega_L
\end{bmatrix}
\begin{bmatrix} \bar{i}_{Sdq} \\ \bar{i}_{LLdq} \\ \bar{u}_{CSdq} \\ \bar{u}_{Ldq} \end{bmatrix} +
\begin{bmatrix} L_S u_{Sdq} \\ 0 \\ 0 \\ 0 \end{bmatrix}
$$

Table 4.7 Mathematical Averaged state-space model of MRPC-II-s

Non-stationary average state-space set equations

$$
\begin{bmatrix} \frac{d\bar{i}_S}{dt} \\ \frac{d\bar{i}_{LL}}{dt} \\ \frac{d\bar{u}_{CS}}{dt} \\ \frac{d\bar{u}_L}{dt} \end{bmatrix}
=
\begin{bmatrix}
-\mathbf{L}_S\mathbf{R}_{LS} & 0 & -\mathbf{L}_S\mathbf{D}_2 & -\mathbf{L}_S\mathbf{D}_M \\
0 & -\mathbf{L}_L\mathbf{R}_{LL} & -\mathbf{L}_L\mathbf{D}_1 & \mathbf{L}_L\mathbf{D}_M \\
\mathbf{C}_S\mathbf{D}_2 & \mathbf{C}_S\mathbf{D}_1 & 0 & 0 \\
\mathbf{C}_L\mathbf{D}_M^T & -\mathbf{C}_L\mathbf{D}_M^T & 0 & -\mathbf{C}_L\mathbf{R}_L
\end{bmatrix}
\begin{bmatrix} \bar{i}_S \\ \bar{i}_{LL} \\ \bar{u}_{CS} \\ \bar{u}_L \end{bmatrix}
+
\begin{bmatrix} \mathbf{L}_S\mathbf{u}_S \\ 0 \\ 0 \\ 0 \end{bmatrix}
$$

Two frequency dq transformation matrix

$$
\mathbf{K} =
\begin{bmatrix}
\mathbf{K}_S & 0 & 0 & 0 \\
0 & \mathbf{K}_S & 0 & 0 \\
0 & 0 & \mathbf{K}_S & 0 \\
0 & 0 & 0 & \mathbf{K}_L
\end{bmatrix}
$$

Stationary average state-space set equations

$$
\begin{bmatrix} \frac{d\bar{i}_{Sdq}}{dt} \\ \frac{d\bar{i}_{LLdq}}{dt} \\ \frac{d\bar{u}_{CSdq}}{dt} \\ \frac{d\bar{u}_{Ldq}}{dt} \end{bmatrix}
=
\begin{bmatrix}
-\mathbf{L}_S\mathbf{R}_{LS} - \Omega_S & 0 & -\mathbf{L}_S\mathbf{D}_2 & -\mathbf{L}_S\mathbf{D}_{Mdq1} \\
0 & -\mathbf{L}_L\mathbf{R}_{LL} - \Omega_S & -\mathbf{L}_L\mathbf{D}_1 & \mathbf{L}_L\mathbf{D}_{Mdq1} \\
\mathbf{C}_S\mathbf{D}_2 & \mathbf{C}_S\mathbf{D}_1 & -\Omega_S & 0 \\
\mathbf{C}_L\mathbf{D}_{Mdq1}^T & -\mathbf{C}_L\mathbf{D}_{Mdq1}^T & 0 & -\mathbf{C}_L\mathbf{R}_L - \Omega_L
\end{bmatrix}
\begin{bmatrix} \bar{i}_{Sdq} \\ \bar{i}_{LLdq} \\ \bar{u}_{CSdq} \\ \bar{u}_{Ldq} \end{bmatrix}
+
\begin{bmatrix} \mathbf{L}_S\mathbf{u}_{Sdq} \\ 0 \\ 0 \\ 0 \end{bmatrix}
$$

Table 4.8 Mathematical Averaged state-space model of MRPC-b

Non-stationary average state-space set equations

$$
\begin{bmatrix} \frac{d\bar{i}_S}{dt} \\ \frac{d\bar{u}_L}{dt} \end{bmatrix}
=
\begin{bmatrix}
-\mathbf{L}_S\mathbf{R}_{LS} & -\mathbf{L}_S\mathbf{D}_M \\
\mathbf{C}_L\mathbf{D}_M^T & -\mathbf{C}_L\mathbf{R}_L
\end{bmatrix}
\begin{bmatrix} \bar{i}_S \\ \bar{u}_L \end{bmatrix}
+
\begin{bmatrix} \mathbf{L}_S\mathbf{u}_S \\ 0 \end{bmatrix}
$$

Two frequency dq transformation matrix

$$
\mathbf{K} =
\begin{bmatrix}
\mathbf{K}_S & 0 \\
0 & \mathbf{K}_L
\end{bmatrix}
$$

Stationary average state-space set equations

$$
\begin{bmatrix} \frac{d\bar{i}_{Sdq}}{dt} \\ \frac{d\bar{u}_{Ldq}}{dt} \end{bmatrix}
=
\begin{bmatrix}
-\mathbf{L}_S\mathbf{R}_{LS} - \Omega_S & -\mathbf{L}_S\mathbf{D}_{Mdq1} \\
\mathbf{C}_L\mathbf{D}_{Mdq1}^T & -\mathbf{C}_L\mathbf{R}_L - \Omega_L
\end{bmatrix}
\begin{bmatrix} \bar{i}_{Sdq} \\ \bar{u}_{Ldq} \end{bmatrix}
+
\begin{bmatrix} \mathbf{L}_S\mathbf{u}_{Sdq} \\ 0 \end{bmatrix}
$$

$$
\begin{aligned}
\bar{i}_{S1} =\ & [U_m \cos(\omega t)(R_L^2 C_F^2 R_{LF}\omega^2(D_S - 1)^4) + R_L(D_S - 1)^2 \\
& \times (q^2 D_S^2(1 + 2R_L C_F C_L R_{LF}\omega\omega_L) + 2C_F^2 R_{LF}\omega^2(R_{LS} - R_L C_L L_S\omega_L^2)) \\
& + (1 + R_L^2 C_L^2\omega_L^2)(q^4 D_S^4 R_{LF} + q^2 D_S^2(R_{LS} - 2C_F L_S R_{LF}\omega\omega_L) \\
& + C_F^2 R_{LF}\omega^2(R_{LS}^2 + L_S^2\omega_L^2)) + U_m \sin(\omega t) R_L^2 C_F\omega(D_S - 1)^4 \\
& \times (C_F L_F\omega^2 - 1) + R_L(D_S - 1)^2(-q^2 D_S^2 R_L C_L\omega_L \\
& + 2C_F L_F R_{LS}\omega^3(C_F - R_{LS}\omega + q^2 R_L C_L D_S^2 L_F\omega^2\omega - R_L C_L L_S\omega\omega_L^2 \\
& \times (C_F L_F\omega^2 - 1))) + (1 + R_L^2 L_L^2\omega_L^2)(q^4 D_S^4 L_F\omega + q^2 D_S^2 L_S\omega_L \\
& \times (1 - 2C_F L_F\omega^2) + C_F\omega(S_F L_F\omega^2 - 1)(R_{LS}^2 + L_S^2\omega_L^2))]/\Delta, \qquad (4.33)
\end{aligned}
$$

Table 4.9 Submatrices connected with non-stationary and stationary models collected in Tables 4.1, 4.2, 4.3, 4.4, 4.5, 4.6, 4.7 and 4.8

Matrices connected with inductances

$$\mathbf{L}_F = \begin{bmatrix} \frac{1}{L_F} & 0 & 0 \\ 0 & \frac{1}{L_F} & 0 \\ 0 & 0 & \frac{1}{L_F} \end{bmatrix}, \ \mathbf{L}_S = \begin{bmatrix} \frac{1}{L_S} & 0 & 0 \\ 0 & \frac{1}{L_S} & 0 \\ 0 & 0 & \frac{1}{L_S} \end{bmatrix}, \ \mathbf{L}_L = \begin{bmatrix} \frac{1}{L_L} & 0 & 0 \\ 0 & \frac{1}{L_L} & 0 \\ 0 & 0 & \frac{1}{L_L} \end{bmatrix}$$

Matrices connected with resistances

$$\mathbf{R}_{LF} = \begin{bmatrix} R_{LF} & 0 & 0 \\ 0 & R_{LF} & 0 \\ 0 & 0 & R_{LF} \end{bmatrix}, \ \mathbf{R}_{LS} = \begin{bmatrix} R_{LS} & 0 & 0 \\ 0 & R_{LS} & 0 \\ 0 & 0 & R_{LS} \end{bmatrix}, \ \mathbf{R}_{LL} = \begin{bmatrix} R_{LL} & 0 & 0 \\ 0 & R_{LL} & 0 \\ 0 & 0 & R_{LL} \end{bmatrix}, \ \mathbf{R}_L = \begin{bmatrix} \frac{1}{R_L} & 0 & 0 \\ 0 & \frac{1}{R_L} & 0 \\ 0 & 0 & \frac{1}{R_L} \end{bmatrix}$$

Matrices connected with capacitances

$$\mathbf{C}_F = \begin{bmatrix} \frac{1}{C_F} & 0 & 0 \\ 0 & \frac{1}{C_F} & 0 \\ 0 & 0 & \frac{1}{C_F} \end{bmatrix}, \ \mathbf{C}_S = \begin{bmatrix} \frac{1}{C_S} & 0 & 0 \\ 0 & \frac{1}{C_S} & 0 \\ 0 & 0 & \frac{1}{C_S} \end{bmatrix}, \ \mathbf{C}_L = \begin{bmatrix} \frac{1}{C_L} & 0 & 0 \\ 0 & \frac{1}{L_L} & 0 \\ 0 & 0 & \frac{1}{C_L} \end{bmatrix}$$

Matrices of duty factors

$$\mathbf{D}_1 = \begin{bmatrix} D_S & 0 & 0 \\ 0 & D_S & 0 \\ 0 & 0 & D_S \end{bmatrix}, \ \mathbf{D}_2 = \begin{bmatrix} (1-D_S) & 0 & 0 \\ 0 & (1-D_S) & 0 \\ 0 & 0 & (1-D_S) \end{bmatrix}, \ \mathbf{D}_M = \begin{bmatrix} d_{aA} & d_{aB} & d_{aC} \\ d_{bA} & d_{bB} & d_{bC} \\ d_{cA} & d_{cB} & d_{cC} \end{bmatrix},$$

$$\mathbf{D}_{Mdq} = \begin{bmatrix} qD_S & 0 & 0 \\ 0 & qD_S & 0 \\ 0 & 0 & D_S \end{bmatrix}, \ \mathbf{D}_{Mdq1} = \begin{bmatrix} q(1-D_S) & 0 & 0 \\ 0 & q(1-D_S) & 0 \\ 0 & 0 & (1-D_S) \end{bmatrix}$$

Voltage vectors

$$\mathbf{u}_S = \begin{bmatrix} u_{S1} \\ u_{S2} \\ u_{S3} \end{bmatrix}, \ \overline{\mathbf{u}}_{CF} = \begin{bmatrix} \overline{u}_{CF1} \\ \overline{u}_{CF2} \\ \overline{u}_{CF3} \end{bmatrix}, \ \overline{\mathbf{u}}_{CS} = \begin{bmatrix} \overline{u}_{CS1} \\ \overline{u}_{CS2} \\ \overline{u}_{CS3} \end{bmatrix}, \ \overline{\mathbf{u}}_L = \begin{bmatrix} \overline{u}_{L1} \\ \overline{u}_{L2} \\ \overline{u}_{L3} \end{bmatrix},$$

$$\mathbf{u}_{Sdq} = \begin{bmatrix} \sqrt{\frac{3}{2}}U_m \\ 0 \\ 0 \end{bmatrix}, \ \overline{\mathbf{u}}_{CFdq} = \begin{bmatrix} \overline{u}_{CF1dq} \\ \overline{u}_{CF2dq} \\ \overline{u}_{CF3dq} \end{bmatrix}, \ \overline{\mathbf{u}}_{CSdq} = \begin{bmatrix} \overline{u}_{CS1dq} \\ \overline{u}_{CS2dq} \\ \overline{u}_{CS3dq} \end{bmatrix}, \ \overline{\mathbf{u}}_{Ldq} = \begin{bmatrix} \overline{u}_{L1dq} \\ \overline{u}_{L2dq} \\ \overline{u}_{L3dq} \end{bmatrix}$$

Current vectors

$$\overline{\mathbf{i}}_S = \begin{bmatrix} \overline{i}_{S1} \\ \overline{i}_{S2} \\ \overline{i}_{S3} \end{bmatrix}, \ \overline{\mathbf{i}}_{LS} = \begin{bmatrix} \overline{i}_{LS1} \\ \overline{i}_{LS2} \\ \overline{i}_{LS3} \end{bmatrix}, \ \overline{\mathbf{i}}_{LL} = \begin{bmatrix} \overline{i}_{LL1} \\ \overline{i}_{LL2} \\ \overline{i}_{LL3} \end{bmatrix}, \ \overline{\mathbf{i}}_{Sq} = \begin{bmatrix} \overline{i}_{S1dq} \\ \overline{i}_{S2dq} \\ \overline{i}_{S3dq} \end{bmatrix}, \ \overline{\mathbf{i}}_{LSdq} = \begin{bmatrix} \overline{i}_{LS1dq} \\ \overline{i}_{LS2dq} \\ \overline{i}_{LS3dq} \end{bmatrix}, \ \overline{\mathbf{i}}_{LLdq} = \begin{bmatrix} \overline{i}_{LL1dq} \\ \overline{i}_{LL2dq} \\ \overline{i}_{LL3dq} \end{bmatrix}$$

$$\begin{aligned}
\overline{u}_{L1} = [&-U_m \cos(\omega_L t) q R_L (D_S - 1) D_S ((D_S - 1)^2 (R_L C_F L_F \omega^2 - R_L) \\
&- (C_F L_S \omega^2 - 1)(R_{LS} + R_L C_L L_S \omega_L^2) + \omega \omega_L (C_F L_L R_{LF} + R_L C_L \\
&\quad (q^2 D_S^2 L_F + C_F R_{LF} R_{LS})) - q^2 D_S^2 R_{LF}) + U_m \sin(\omega_L t) q R_L (D_S - 1) \\
&\times D_S (R_L C_F (D_S - 1)^2 R_{LF} \omega + L_S \omega_L + R_L C_L \omega_L (q^2 D_S^2 R_{LF} + R_{LS} \\
&- C_F L_F R_{LS} \omega^2) + \omega (q^2 D_S^2 L_F + C + F R_{LF} R_{LS} - C_F L_S \omega_L \\
&\times (L_F \omega + R_L C_L R_{LF} \omega_L)))]/,
\end{aligned} \quad (4.34)$$

$$
\begin{aligned}
\Delta = {}& R_L^2 (D_S - 1)^4 (1 + C_F \omega^2 (C_F R_{LF}^2 + L_F (C_F L_F \omega^2 - 2))) + 2R_L (D_S - 1)^2 \\
& \times (q^2 D_S^2 (R_{LF} + R_L C_L \omega \omega_L (C_F R_{LF} + L_F (C_F L_F \omega^2 - 1))) \\
& + (1 + C_F \omega^2 (C_F R_{LF}^2 + L_F (C_F L_F \omega^2 - 2)))(R_{LS} - R_L C_L L_S \omega_L^2)) \\
& + (1 + R_L^2 C_L^2 \omega_L^2)[q^4 D_S^4 (R_{LF}^2 + L_L^2 \omega^2) + 2q^2 D_S^2 (R_{LF} R_{LS} - L_S \omega \omega_L \\
& \times (C_F R_{LF}^2 + L_F (C_F L_L \omega^2 - 1))) + (1 + C_F \omega^2 (C_F R_{LF}^2 + L_F \\
& \times (C_F L_F \omega^2 - 2)))(R_{LS}^2 + L_S^2 \omega_L^2)].
\end{aligned}
\tag{4.35}
$$

4.6 Chapter Summary

The modelling approach based on the averaged state-space method presented in this chapter is relatively simple and requires only a small number of mathematical transformations. Additional circuit models (equivalent schemes, one phased basic schemes, signal flow graphs) are not required for this method. The average set equation is obtained directly from the three-phase schematic circuit, taking into account the sequences of switching pattern and modulation strategies. For the purpose of obtaining stationary equations that are independent of time a two-frequency dq transformation is used. Equations for both steady and transient states obtained in this way have solutions that are well known in the literature. The difficulty in this description is the quite complicated analytic solutions. Currently, the analytical or numerical method of solving this problem is simple using advanced mathematical software.

References

1. Chen J, Ngo DT (2001) Graphical phasor analysis of three-phase PWM converters. IEEE Trans Power Electron 16(5):659–666
2. Fedyczak Z (2003) PWM AC voltage transforming circuits (in Polish). Zielona Góra University Press, Zielona Góra
3. Fedyczak Z, Szcześniak P (2006) Koncepcja matrycowo-reaktancyjnego przemiennika częstotliwości typu Ćuk (in Polish). Przegląd Elektrotechniczny (Electr Rev) 7/8:42–47
4. Fedyczak Z, Szcześniak P (2006) Koncepcja matrycowo-reaktancyjnego przemiennika częstotliwości typu Zeta (in Polish). Wiadomości Elektrotechniczne (Electrotech News) 3:26–29
5. Fedyczak Z, Szcześniak P (2012) Matrix-reactance frequency converters using an low frequency transfer matrix modulation method. Electr Power Syst Res 83(1):91–103
6. Fedyczak F, Szcześniak P (2009) Modelling and analysis of matrix-reactance frequency converters using voltage source matrix converter and LF transfer matrix modulation method. Przegląd Elektrotechniczny (Electr Rev) 2:125–130
7. Fedyczak Z, Szcześniak P (2007) New matrix-reactance frequency converters-conception description. In: Orłowska-Kowalska T (ed) Power electronics and electrical drives: selected problems. Wrocław Technical University Press, Wrocław, pp 71–84

8. Fedyczak, Z, Szcześniak P (2005) Study of matrix-reactance frequency converter with buck-boost topology. In: Proceedings of power electronics and intelligent control for energy conservation conference, PELINCEC'05, Warsaw, Poland (CD-ROM)
9. Fedyczak Z, Klytta M, Strzelecki R (2001) Three-phase AC/AC semiconductor transformer topologies and applications. In: Proceedings of power electronics devices compatibility conference, PEDC'01, Zielona Góra, Poland, pp 25–38
10. Fedyczak Z, Strzelecki R, Sozański K (2002) Review of three-phase AC/AC semiconductor transformer topologies and applications. In: Proceedings of symposium power electronics electrical drives automation and motion, SPEEDAM'02, Ravello, Italy, pp B.5-19–B.5-24
11. Fedyczak Z, Szcześniak P, Jankowski M (2005) Koncepcja matrycowo-reaktacyjnego przemiennika częstotliwości typu buck-bost (in Polish). Sterowanie w Energoelektronice i Napędzie Elektrycznym, SENE'05, number 1, Łódź, Poland, pp 101–106
12. Fedyczak Z, Szcześniak P, Kaniweski J (2007) Direct PWM AC choppers and frequency converters. In: Korbicz J (ed) Measurements models systems and design. Transport and Communication Publishers, Warsaw, pp 393–424
13. Fedyczak Z, Szcześniak P, Klytta M (2006) Matrix-reactance frequency converter based on buck-boost topology. In: Proceedings of power electronics and motion control conference, EPE-PEMC'06, Portoroz, Slovenia, pp 763–768
14. Fedyczak Z, Szcześniak P, Korotyeyev I (2008) Generation of matrix-reactance frequency converters based on unipolar PWM AC matrix-reactance choppers. In: Proceedings of IEEE power electronics specialists conference, PESC'08, Rhodes, Greece, pp 1821–1827
15. Fedyczak Z, Szcześniak P, Korotyeyev I (2008) New family of matrix-reactance frequency converters based on unipolar PWM AC matrix-reactance choppers. In: Proceedings of power electronics and motion control conference, EPE-PEMC'08, Poznań, Poland, pp 236–243
16. Fedyczak Z, Szcześniak P, Kaniweski J, Tadra G (2009) Implementation of three-phase frequency converters based on PWM AC matrix-reactance chopper with buck-boost topology. In: Proceedings of European conference on power electronics and applications, EPE'09, Barcelona, Spain, pp P1–P10 (CD-ROM)
17. Fedyczak Z, Tadra G, Klytta M (2010) Implementation of the current source matrix converter with space vector modulation. In: Proceedings of power electronics and motion control conference, EPE-PEMC'10, Ohrid, Macedonia (CD-ROM)
18. Gao F, Iravani MR (2007) Dynamic model of a space vector modulated matrix converter. IEEE Trans Power Deliv 22(3):1696–1705
19. Kanaan HY, Al-Hadad K (2003) A new average modeling and control design applied to a nine-switch matrix converter with input power factor correction. In: Proceedings of EPE'03, Toulouse, France (CD-ROM)
20. Kanaan HY, Al-Hadad K (2002) A comparison between three modeling approaches for computer implementation of high-fixed-switching-frequency power converters operating in a continuous mode. In: Proceedings of Canadian conference on electrical and computer engineering CCECE'02, Winnipeg, Canada, pp 12–15
21. Korotyeyev I, Fedyczak Z (2008) Steady and transient states modelling methods of matrix-reactance frequency converter with buck-boost topology. COMPEL: Int J Comput Math Electr Electron Eng 28(3):626–638
22. Korotyeyev I, Fedyczak Z, Szcześniak P (2008) Steady and transient state analysis of a matrix-reactance frequency converter based on a boost PWM AC matrix-reactance chopper. In: Proceedings of the international school on nonsinusoidal currents and compensation, ISNCC'08, Łagów, Poland (CD-ROM)
23. Korotyeyev I, Fedyczak Z, Strzelecki R, Sozański KP (2001) An averaged AC models accuracy evaluation of non-isolated matrix-reactance PWM AC line conditioners. In: Proceedings of European conference on power electronics and applications, EPE'01, Graz (CD-ROM)
24. Krein PT, Bentaman J, Bass RM, Lesieutre BC (1990) On the use of averaging for the analysis of power electronic systems. IEEE Trans Power Electron 5(2):182–190
25. Kwon WH, Cho GH (1993) Analyses of static and dynamic characteristics of practical step-up nine-switch convertor. IEE Proc-B 140(2):139–145

26. Kwon WH, Cho GH (1991) Analysis of non-ideal step down matrix converter based on circuit DQ transformation. In: Proceedings of power electronics specialists conference, PESC'91, Cambridge, US, pp 825–829
27. Middlebrook RD, Ćuk S (1976) A general unified approach to modelling switching-converter power stages. In: Proceedings of power electronics specialists conference, PESC'76, Cleveland, US, pp 73–86
28. Rim CT, Hu DY, Cho GH (1990) Transformers as equivalent circuits for switches: general proofs and D-Q transformation-based analyses. IEEE Trans Ind Appl 26(4):777–785
29. Rim CT (2011) Unified general phasor transformation for AC converters. IEEE Trans Power Electron 26(9):2465–2475
30. Szcześniak P (2009) Analysis and testing matrix-reactance frequency converters, PhD thesis (in Polish), University of Zielona Góra, Zielona Góra
31. Szcześniak P (2007) Basic properties comparative study of matrix-reactance frequency converter based on buck-boost topology with Venturini control strategies. In: Proceedings of compatibility in power electronics, CPE'07, Gdańsk, Poland (CD-ROM)
32. Szcześniak P (2010) Modele matematyczne trójfazowych przemienników częstotliwości prądu przemiennego bazujących na topologii sterownika matrycowo-reaktancyjnego typu buck-boost (in Polish). Przegląd Elektrotechniczny (Electr Rev) 2:384–389
33. Szcześniak P, Fedyczak Z, Klytta M (2008) Modelling and analysis of a matrix-reactance frequency converter based on buck-boost topology by DQ0 transformation. In: Proceedings of power electronics and motion control conference, EPE-PEMC'08, Poznań, Poland, pp 165–172
34. Szcześniak P, Fedyczak Z, Tadra G (2011) Modeling of the matrix-reactance frequency converters using SVM method (in Polish). In: Proceedings of Sterowanie w Energoelektronice i Napędzie Elektrycznym, SENE 2011, Łódź, Poland (CD-ROM)

Chapter 5
Property Analysis

5.1 Introduction

In Chap. 4 the topic of matrix-reactance frequency converters mathematical modelling is discussed. The construction of a mathematical model is described, and analytical expressions are derived and evaluated. Using the mathematical description given by Eq. (4.24) the steady-state characteristics and time waveforms are obtained, whereas the transient states time waveforms are obtained from expression (4.23). The purpose of this chapter is to present these analytical results in easily understandable and readily usable graphical and tabular forms. The main objective is to provide a general appreciation of the practical significance of, and differences between, the external operating characteristics of the various MRFCs.

In this chapter, matrix-reactance frequency converters with balanced three-phase supply and load are considered. Also, it is assumed that the AC source has zero internal impedance. The circuit parameters are shown in Table 5.1. The properties of all topologies were examined for the same circuit parameters. Due to the large number of topologies, detailed extended results were shown only for the selected topology of MRFC-I-b-b. For this topology, the static characteristics for various load conditions and control parameters have been plotted. Furthermore, only for this topology is the transient state analysis presented. For other topologies, the basic static characteristics have been drawn. In the final part of this chapter, simulation studies of MRPC-I-b-b in the drive system with a cage asynchronous motor are presented. At the end of the chapter, a summary of the MRFCs properties, in tabular form and overall characteristics are presented. The practical implementation guidelines for the design of MRFCs are also depicted.

P. Szcześniak, *Three-Phase AC–AC Power Converters Based on Matrix Converter Topology*, Power Systems, DOI: 10.1007/978-1-4471-4896-8_5,
© Springer-Verlag London 2013

Table 5.1 Circuits parameters

Parameter	Symbol	Values
Supply voltage	U_S	230 V
Sfrequency of supply voltage	f	50 Hz
Source filter inductance	$L_{F1}, L_{F2}, L_{F3}, L_F$	1.5 mH
Input inductance	$L_{S1}, L_{S2}, L_{S3}, L_S$	1.5 mH
Output inductance	$L_{L1}, L_{L2}, L_{L3}, L_L$	1.5 mH
Resistance of source filter inductors	$R_{LF1}, R_{LF2}, R_{LF3}, R_{LF}$	0.01 Ω
Resistance of input inductors	$R_{LS1}, R_{LS2}, R_{LS3}, R_{LS}$	0.01 Ω
Resistance of output inductors	$R_{LL1}, R_{LL2}, R_{LL3}, R_{LL}$	0.01 Ω
Source filter capacitances	$C_{F1}, C_{F2}, C_{F3}, C_F$	10 μF
Input capacitances	$C_{S1}, C_{S2}, C_{S3}, C_S$	10 μF
Output capacitances	$C_{L1}, C_{L2}, C_{L3}, C_L$	10 μF
Resistance of loads	$R_{L1}, R_{L2}, R_{L3}, R_L$	60 Ω

5.2 Steady-State Analysis

Investigations that show the properties of the matrix-reactance frequency family of converters were carried out in a system consisting of the resistance load and the idealised power stage elements. Detailed results are shown only for MRFC-I-b-b topology (Fig. 3.7). The first step in the analysis of this circuit is to obtain the averaged output voltage waveforms (u_{L1}) for different setting frequencies $f_L = 25, 50$ and 75 Hz, that describe basic power-stage circuit operation—regulation of output voltage amplitude. These time waveforms are compiled in Fig. 5.1a, c, e. The waveforms of the output voltages were obtained using straightforward theoretical analysis techniques, based on the solution of Eq. (4.24), which for the analysed topology is given by (4.34) and (4.35). The presented output voltage waveforms are juxtaposed with source voltage u_{L1} time waveforms. The time waveforms shown in Fig. 5.1 confirm that by means of the discussed MRFC-I-b-b circuit frequency conversion and buck-boost load voltage changes are possible. For the $D_S = 0.75$ amplitude of output voltage u_{L1} is greater than the amplitude of source voltage u_{S1}. This theoretical output voltage agrees well with that obtained from real-time PSpice simulation, as is evident in Fig. 5.1b, d, f. Generally the results of the simulation experiment confirm the results of theoretical studies. Differences between analytic and simulation results are caused by higher harmonics being taking into account during the simulation experiment (non-stationary circuit).

To illustrate the effect of the discussed output voltage control, typical characteristics are shown with various sequence pulse duty factor D_S and setting output voltage frequency. Shown in Fig. 5.2 are the characteristics of voltage and current gain and input power factor as functions of load voltage setting frequency and pulse duty factor D_S, obtained by means of (4.24) and (4.33)–(4.35) for circuit parameters collected in Table 5.1. From these characteristics it is also visible that both a frequency conversion and a buck-boost load voltage change are possible. Using a simple control

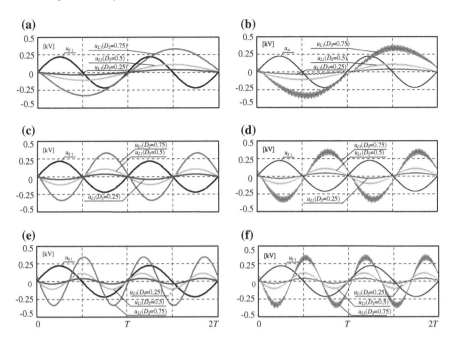

Fig. 5.1 Steady-state output voltage u_{L1} of MRFC-I-b-b for different sequence pulse duty factor D_S and setting output frequency: obtained by theoretical analysis **a** $f_L = 25\,\text{Hz}$, **c** $f_L = 50\,\text{Hz}$, **e** $f_L = 75\,\text{Hz}$: obtained by simulation, **b** $f_L = 25$ Hz, **d** $f_L = 50\,\text{Hz}$, **f** $f_L = 75\,\text{Hz}$

strategy, attributable to Venturini [11], for sequence pulse duty factor $D_S > c.a.$ 0.65 a load voltage greater than supply voltage can be obtained.

It can be seen from Fig. 5.2 that the frequency of output voltage has a significant effect on the converter properties. This chart clearly illustrates that the amplitude of output voltage for lower output frequency f_L is greater than the amplitude for the higher frequency (Fig. 5.2a). A similar influence of output setting frequency is visible on the characteristic of current gain (Fig. 5.2b) and input power factor λ_p (Fig. 5.2c). These visible differences are caused by the influence of the passive element parameters, which are used in the MRFC circuit. The MRFCs topologies are resonant RLC circuits. The influence of output voltage setting frequency on MRFC properties is presented in 3D form shown in Fig. 5.3.

The MRFC-I-b-b current input and output relations are nonlinear and strongly depend on the D_S factor, as shown in Figs. 5.2b and 5.3b. For the higher value of D_S (greater than 0.8) the input currents (i_{S1}, i_{S2}, i_{S3}) and source inductor currents ($i_{LS1}, i_{LS2}, i_{LS3}$) are much higher than for the lower value of D_S. The sample characteristic of current amplitudes in the discussed topology are presented in Fig. 5.4. By increasing the D_S ratio above 0.8, we take a large current from the source. Also, the amplitude of source inductor current is large. Inspection of Fig. 5.4 indicates that application of the discussed MRFC with high D_S is disadvantageous. In this case the magnetic saturation of the inductor cores can occur. The long-term magnetic

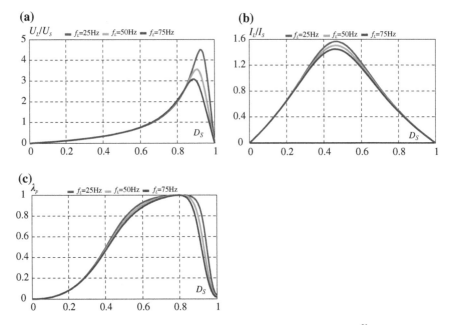

Fig. 5.2 Steady-state characteristics of MRFC-I-b-b: **a** voltage gain $K_U = \frac{U_L}{U_S}$, **b** current gain $K_I = \frac{I_L}{I_S}$, **c** input power factor λ_P

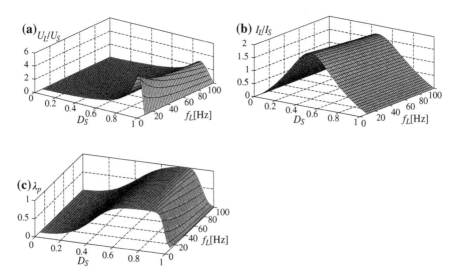

Fig. 5.3 Influence of load frequency on MRFC-I-b-b properties: **a** voltage gain K_U, **b** current gain K_I, **c** input power factor λ_P

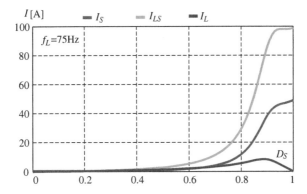

Fig. 5.4 Steady-state characteristic of current amplitudes in MRFC-I-b-b for $f_L = 25\,\mathrm{Hz}$

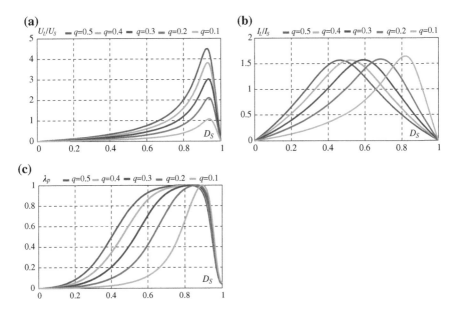

Fig. 5.5 Influence of coefficient q on MRFC-I-b-b properties for $f_L = 25\,\mathrm{Hz}$: **a** voltage gain K_U, **b** current gain K_I, **c** input power factor λ_P

saturation of the cores usually leads to inductor or converter damage. The level of the inductor current must be taken into consideration in the design process.

As explained above, the output voltage waveform regulation is composed of two segments. The first is connected with the operation of matrix connected switches (S_{jK}) and the second with additional switches S_{L1}, S_{L2}, S_{L3}. In MRFC topologies there are two levels of output voltage regulation. There are additional degrees of control freedom. The first was presented previously and concerns modulation of sequence duty pulse D_S (Figs. 5.1, 5.2, 5.3 and 5.4). The second degree of control

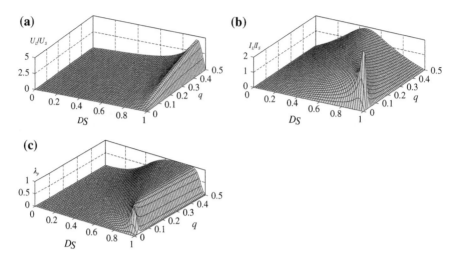

Fig. 5.6 Steady-state 3D characteristics of MRFC-I-b-b: **a** voltage gain K_U, **b** current gain K_I, **c** input power factor λ_P

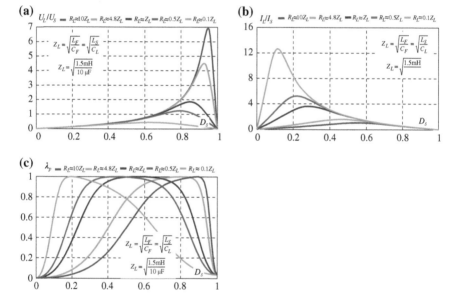

Fig. 5.7 Influence of load condition on MRFC-I-b-b properties for $f_L = 25\,\text{Hz}$: **a** voltage gain K_U, **b** current gain K_I, **c** input power factor λ_P

freedom of output voltage in MRFC is given by the variable of matrix voltage gain q. Coefficient q is defined as a setting quotient of output and input voltages on matrix switches. All time waveforms and static characteristic from Figs. 5.1, 5.2, 5.3 and 5.4 are presented with optimum value of coefficient $q = 0.5$ (optimum for Venturini control strategy). Figure 5.5 shows the variation in the MRFC-I-b-b

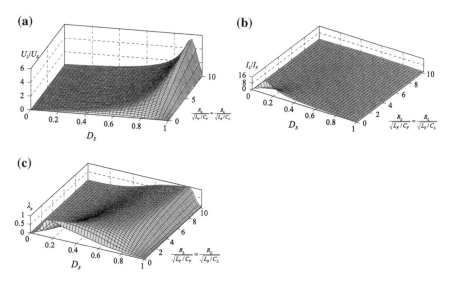

Fig. 5.8 Influence of condition load on MRFC-I-b-b properties for $f_L = 25$ -3D characteristics: **a** voltage gain K_U, **b** current gain K_I, **c** input power factor λ_P

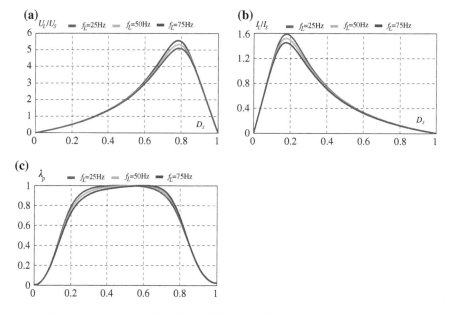

Fig. 5.9 Steady-state characteristics of MRFC-II-b-b: **a** voltage gain K_U, **b** current gain K_I, **c** input power factor λ_P

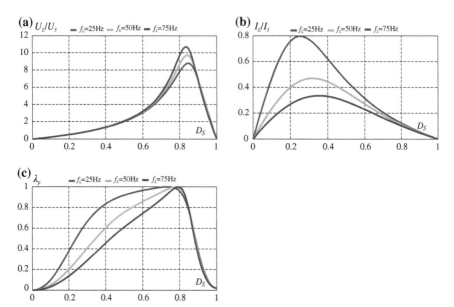

Fig. 5.10 Steady-state characteristics of MRFC-I-c: **a** voltage gain K_U, **b** current gain K_I, **c** input power factor λ_P

properties for a different value of q as a function of sequence pulse duty factor D_S. All the possibilities of output voltage control are presented in 3D characteristics in Fig. 5.6. In this figure, the two levels of control with q and D_S variations are presented. The additional degree of control freedom provided by the coefficient q can be beneficial for MRFCs control properties.

The MRFC is equivalent to an *RLC* resonant circuit whose values are varied by the load character. Furthermore, the MRFC transfers energy from sources to the load, along with their properties, such as in transmission lines. These types of systems are characterised by characteristic impedance Z_{Ch}. Thus, maximum transfer of power from the source to the load takes place when the load is matched to the source. The parameters of MRFC-I-b-b circuits and characteristic impedance Z_{Ch} are related as follows [1]:

$$Z_{Ch} = \sqrt{\frac{L_F}{C_F}} = \sqrt{\frac{L_S}{C_L}}. \tag{5.1}$$

The full load-matching condition is achievable if:

$$Z_{Ch} = \underline{Z}_L = R_L. \tag{5.2}$$

Efficient power transfer is possible with other converter and load impedances, but with less capacity. In addition to the problem of matching a purely resistive load that is not equal to Z_{Ch}, a typical practical problem includes substantial load reactance as well. Figure 5.7 shows the influence of load conditions on MRFC-I-b-b

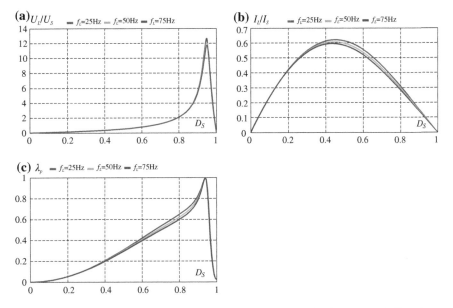

Fig. 5.11 Steady-state characteristics of MRFC-II-c: **a** voltage gain K_U, **b** current gain K_I, **c** input power factor λ_P

properties. The analysed MRFC is sensitive to changes in the load parameters. In this book only the resistance load condition is analysed. For load resistance equal to or greater than the characteristic impedance ($R_L \geq Z_{Ch}$) a voltage transfer ratio greater than one is achieved. In the case when $R_L \leq Z_{Ch}$, the output voltage is less than source voltage. Thus in the load-matched case there are minimum power losses and maximum efficiency. The 3D chart of Fig. 5.8 shows in detail the analysis converter's sensitivity of parameters to load conditions.

All the results presented in Figs. 5.7 and 5.8 show the limits of load variation. These limits are the reason that MRFCs have a potential application as a universal frequency converter. The relationship of MRFCs LC components and load conditions should be individually selected depending on the type of load.

The presented steady-state time waveforms and characteristics concern the MRFC-I-b-b topology. Other topologies of MRFCs have similar properties to the MRFC-I-b-b. The setting load frequency and load conditions, too, have an influence on output voltages and the input power factor. Similarly, two degrees of control freedom are given, with variations of q and D_S coefficients. However, the static characteristics are different. For the other topologies only basic static characteristics are indicated, showing the relations of voltage, current gain and input power factor as a function of D_S for three output frequencies $f_L - 25, 50, 75$ Hz.

Analytical characteristics for MRFC-II-b-b were obtained using theoretical analysis techniques, based on the solution of Eq. (4.24), in the form presented in Table 4.1. Figure 5.9 illustrates the obtained results. It can be seen that the obtained voltage

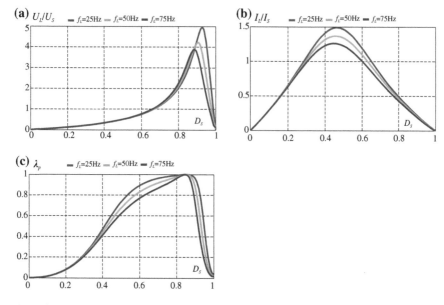

Fig. 5.12 Steady state characteristics of MRFC-I-z: **a** voltage gain K_U, **b** current gain K_I, **c** input power factor λ_P

gain for MRFC-II-b-b is greater than the voltage gain for MRFC-I-b-b. Also, the performance of the input power factor is better. In practical applications, when voltage gain of a little larger than one is needed a detailed analysis and design procedure is required. In this topology smaller capacitances and inductor can be applied.

Similar characteristics are presented for MRFC-I-c (Fig. 5.10) and MRFC-II-c (Fig. 5.11). The average stationary equations are included in the Tables 4.2 and 4.3 for MRFC-I-c and MRFC-II-c, respectively. To illustrate the properties of Zeta MRFC topologies similar results are presented in Figs. 5.12 and 5.13 for MRFC-I-z and MRFC-II-z, respectively. The mathematical models for these topologies are presented in Tables 4.4 and 4.5.

The graphs in Figs. 5.14 and 5.15 correspond to steady-state operation in SEPIC MRFC topologies, and were obtained from expressions included in Tables 4.6 and 4.7. The last plot in Fig. 5.16 shows the MRFC-b properties in a steady state. The boost MRFC averaged model is presented in Table 4.8.

The concluding part of the chapter is concerned with the summary of properties of all MRFCs. The overall characteristics and tabular comparison are presented. The practical implementation guidelines for the design of MRFCs are also outlined in the summary of the chapter.

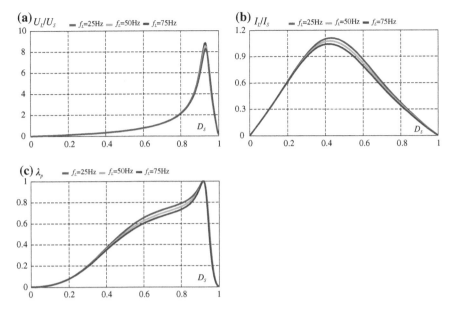

Fig. 5.13 Steady-state characteristics of MRFC-II-z: **a** voltage gain K_U, **b** current gain K_I, **c** input power factor λ_P

5.3 Transient State Analysis

The performance of an MRFC is studied in terms of steady-state and transient-state analysis. The steady-state analysis is important for selecting the converter operating conditions, ensuring the optimum working conditions for a particular application and for a given configuration of load and power supply. The transient-state analysis of an MRFC is vital for assessing the dynamics of the system and disturbances generated during transients as well as to develop a protection strategy. The transient simulations presented in the literature are used in two main groups of analyses: transient analysis for step changes of control signals and power stage parameters. The models developed for transient stability analysis are generally valid in longer time frames.

These studies are focused on the converter analysis of step change in control signals, such as the sequence pulse duty factor D_S and setting load frequency f_L. Furthermore, the supply voltage changes will be analysed. Using the mathematical description given by Eq. (4.23) the transient state time waveforms are obtained. Similarly, as in the above subchapter only MRFC-I-b-b topology was analysed.

In Figs. 5.17 and 5.18 are shown exemplary voltage and current time waveforms illustrating the selected cases of the transient states in the discussed MRFC. Figure 5.17 depicts the theoretical transient responses of state variables at a step change of the sequence pulse duty factor D_S from 0.5 to 0.7 for output frequencies $f_L = 25$ Hz. The responses of state variables at a step change in the supply voltages

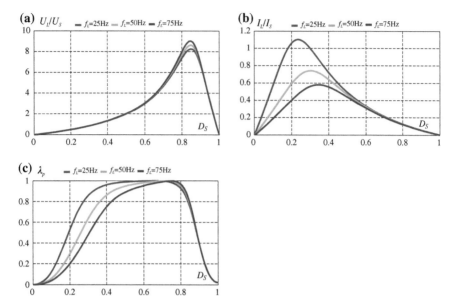

Fig. 5.14 Steady-state characteristics of MRFC-I-s: **a** voltage gain K_U, **b** current gain K_I, **c** input power factor λ_P

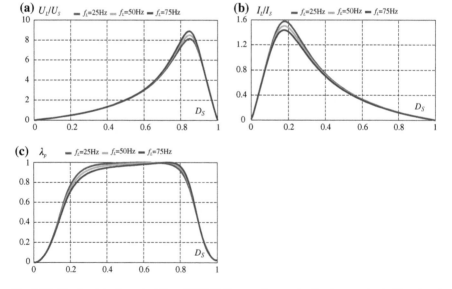

Fig. 5.15 Steady-state characteristics of MRFC-II-s: **a** voltage gain K_U, **b** current gain K_I, **c** input power factor λ_P

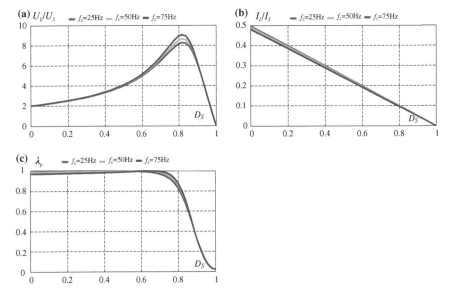

Fig. 5.16 Steady-state characteristics of MRFC-b: **a** voltage gain K_U, **b** current gain K_I, **c** input power factor λ_P

U_{Sm} from 150 to 230 V for output frequencies $f_L = 25$ Hz at $D_S = 0.75$, can be observed in Fig. 5.18.

The waveforms in Figs. 5.17 and 5.18 confirm good dynamic properties of such a converter. The converter transient response is relatively short. In the case of a step change in coefficient D_S, the transient period is approximately equal to 0.25 times the supply voltage period, but the current distortion has a longer time response. There are high frequency oscillations which last approximately to three periods of this current, and afterwards the system attains steady-state condition. In the second case the time response is also short. The main distortion lasts about 0.5 times to period of source voltage, but the high frequency oscillations last more than 4 periods.

To verify the averaged modelling accuracy, Fig. 5.19 shows the state variables' waveforms during the transience. The waveforms obtained by switched-circuit transient PSpice simulation are shown together with the waveforms obtained by analysis of the averaged model (4.23). As shown in Fig. 5.19 the calculation and simulation test results demonstrate good correlation which confirms the usefulness of the used analytical averaged method.

The properties of other MRFCs in transient state are not presented in this book. Some results can be found in [2–5, 8, 9].

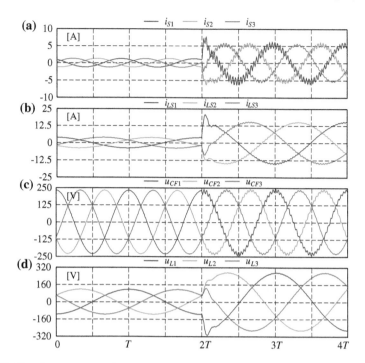

Fig. 5.17 Transient responses of states variables at step change in the sequence pulse duty factor D_S from 0.5 to 0.7, for $f_L = 25\,\text{Hz}$

5.4 Drive System Application

As an exemplary application of the MRFC devices shown above an application in a power electric drive system with induction cage motor is shown (Fig. 5.20). Simulation of a drive system with an MRFC and cage asynchronous motor has been carried out with the help of the PSpice computer program. A cage motor simulation model was constructed in PSpice based on mathematical relations (5.3)–(5.14) [6], which describe the motor operation in a coordinate system rotating with an angular speed ω, called the coordinate system $\alpha\beta\gamma$.

$$u_{s\alpha} = R_s i_{s\alpha} + \frac{\omega(\psi_{s\gamma} - \psi_{s\beta})}{\sqrt{3}} + \frac{d\psi_{s\alpha}}{dt}, \qquad (5.3)$$

$$u_{s\beta} = R_s i_{s\beta} + \frac{\omega(\psi_{s\alpha} - \psi_{s\gamma})}{\sqrt{3}} + \frac{d\psi_{s\beta}}{dt}, \qquad (5.4)$$

$$u_{s\gamma} = R_s i_{s\gamma} + \frac{\omega(\psi_{s\beta} - \psi_{s\alpha})}{\sqrt{3}} + \frac{d\psi_{s\gamma}}{dt}, \qquad (5.5)$$

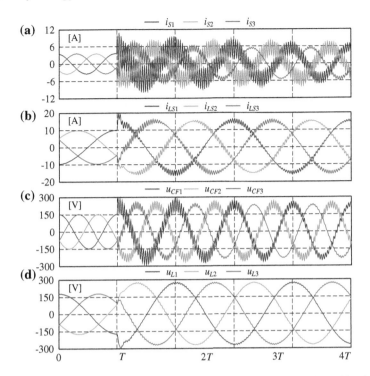

Fig. 5.18 Transient responses of state variables at step change in the supply voltage U_{Sm} from 150 to 230 V, for $f_L = 25\,\text{Hz}$ at $D_S = 0.75$

$$u'_{r\alpha} = R'_r i'_{r\alpha} + (\omega - \omega_r)\frac{(\psi'_{r\gamma} - \psi'_{r\beta})}{\sqrt{3}} + \frac{d\psi'_{r\alpha}}{dt}, \tag{5.6}$$

$$u'_{r\beta} = R'_r i'_{r\beta} + (\omega - \omega_r)\frac{(\psi'_{r\alpha} - \psi'_{r\gamma})}{\sqrt{3}} + \frac{d\psi'_{r\beta}}{dt}, \tag{5.7}$$

$$u'_{r\gamma} = R'_r i'_{r\gamma} + (\omega - \omega_r)\frac{(\psi'_{r\beta} - \psi'_{r\alpha})}{\sqrt{3}} + \frac{d\psi'_{r\gamma}}{dt}, \tag{5.8}$$

$$\psi_{s\alpha} = L_{sl}i_{s\alpha} + M(i_{s\alpha} - i'_{r\alpha}), \tag{5.9}$$

$$\psi_{s\beta} = L_{sl}i_{s\beta} + M(i_{s\beta} - i'_{r\beta}), \tag{5.10}$$

$$\psi_{s\gamma} = L_{sl}i_{s\gamma} + M(i_{s\gamma} - i'_{r\gamma}), \tag{5.11}$$

$$\psi'_{r\alpha} = L'_{rl}i'_{r\alpha} + M(i_{s\alpha} - i'_{r\alpha}), \tag{5.12}$$

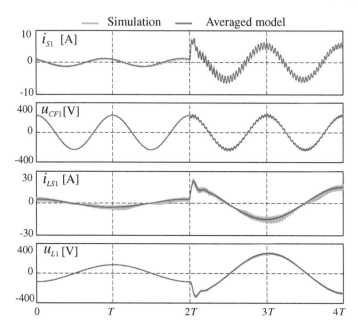

Fig. 5.19 Comparison of PSpice simulation and averaged model results from transient responses of state variables at step change in the sequence pulse duty factor D_S from 0.5 to 0.7, $f_L = 25\,\text{Hz}$

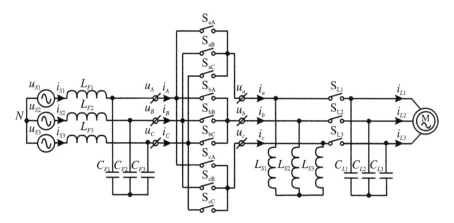

Fig. 5.20 Drive system with MRFC-I-b-b and induction cage motor

$$\psi'_{r\beta} = L'_{sl}i'_{r\beta} + M(i_{s\beta} - i'_{r\beta}), \tag{5.13}$$

$$\psi'_{r\gamma} = L'_{rl}i'_{r\gamma} + M(i_{s\gamma} - i'_{r\gamma}), \tag{5.14}$$

where: $u_{s\alpha}$, $u_{s\beta}$, $u_{s\gamma}$, $i_{s\alpha}$, $i_{s\beta}$, $i_{s\gamma}$—stator voltages and currents that are described in $\alpha\beta\gamma$ coordinates rotating with an angular speed ω; $u'_{r\alpha}$, $u'_{r\beta}$, $u'_{r\gamma}$, $i'_{r\alpha}$, $i'_{r\beta}$, $i'_{r\gamma}$ - rotor

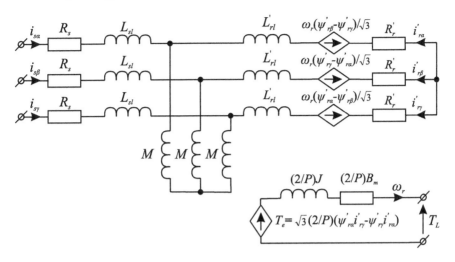

Fig. 5.21 Simplified model of cage induction motor described in $\alpha\beta\gamma$ coordinate

voltages and currents that are described in $\alpha\beta\gamma$ coordinates rotating with an angular speed ω which are reduced to stator coordinate; R_s, R_r—stator and rotor resistances; L_{sl}, L_{rl}—leakage inductance of stator and rotor windings; $M = \frac{3}{2}L_{ms}$—magnetising inductance; L_{ms}—magnetising inductance of the stator winding; $\psi_{s\alpha}$, $\psi_{s\beta}$, $\psi_{s\gamma}$—rotor flux described in $\alpha\beta\gamma$ coordinates; $\psi'_{r\alpha}$, $\psi'_{r\beta}$, $\psi'_{r\gamma}$—stator flux described in $\alpha\beta\gamma$ coordinates which are reduced to stator coordinates.

The mechanical system is described by the following equation [6]:

$$T_e = \frac{2}{P}J\omega_r + \frac{2}{P}B_m\omega_r + T_L, \qquad (5.15)$$

where: T_e—electromagnetic torque; P—number of pole pairs; J—moment of inertia of the mechanical system; ω_r—rotor angular speed; T_L—load moment; B_m—viscous friction coefficient. Electromagnetic torque can also be represented by the following equation [6]:

$$T_e = \sqrt{3}\frac{P}{2}(\psi'_{r\alpha}i'_{r\gamma} - \psi'_{r\gamma}i'_{r\alpha}) \qquad (5.16)$$

Equations (5.15) and (5.16) are implemented in the simulation model to approximate the mechanical parameters of the system and the loading conditions. For the induction machine fed from the stator circuit the rotation speed in $\alpha\beta\gamma$ coordinate is set by zero $\omega = 0$. Then, a simplified model of the cage induction motor determined by Eqs. (5.3)–(5.16) are presented in Fig. 5.21 [6].

A computer simulation was carried out using the simulation parameters of the inverter and the motor put together in Table 5.2 [7, 8]. The simulation results of a drive system with MRFC-I-b-b topology can be observed in 5.22–5.24. In Figs. 5.22 and 5.23 are presented the exemplary time waveforms of converter input and output

Table 5.2 Simulation parameters of drive system with MRFC-I-b-b

Parameter	Symbol	Value
Supply voltage	U_S	230 V
Frequency of supply voltage	f	50 Hz
Switching frequency	f_S	5 kHz
Maximum simulation step	t_p	1 μs
Source filter inductance	L_{F1}, L_{F2}, L_{F3}	0.5 mH
Source inductance	L_{S1}, L_{S2}, L_{S3}	0.5 mH
Source filter capacitance	C_{F1}, C_{F2}, C_{F3}	20 μF
Load capacitance	C_{L1}, C_{L2}, C_{L3}	20 μF
Motor power	P_n	2.2 kW
Switch resistance in turn-on state	R_{ON}	0.01 Ω
Switch resistance in turn-off state	R_{OFF}	0.1 MΩ
Stator and rotor resistance	R_s, R_r'	2.5002 Ω
Leakage inductance of stator and rotor windings	L_{sl}, L_{rl}'	0.011 H
Magnetizing inductance	M	0.4576 H
Moment of inertia	J	0.06825 kg m^2
Viscous friction coefficient	B_m	1.024 10^{-3}
Number of pole pairs	P	4

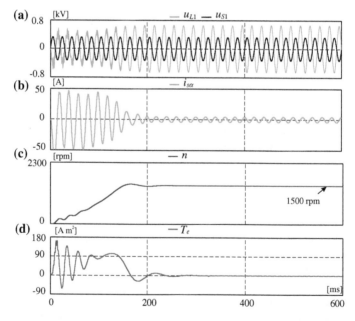

Fig. 5.22 Simulation time waveforms of drive system with MRFC-I-b-b during motor starting, for $f_L = 50$ Hz, $D_S = 0.8$, $q = 0.5$: **a** input (u_{S1}) and output (u_{L1}) voltage of converter, **b** motor current ($i_{S\alpha}$), **c** speed (n), **d** electromagnetic torque (T_e)

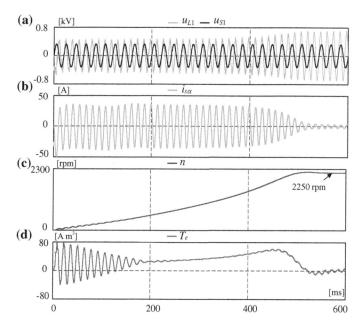

Fig. 5.23 Simulation time waveforms of drive system with MRFC-I-b-b during motor starting, for $f_L = 75\,\text{Hz}$, $D_S = 0.8$, $q = 0.5$: **a** input (u_{S1}) and output (u_{L1}) voltage of converter, **b** motor current ($i_{S\alpha}$), **c** speed (n), **d** electromagnetic torque (T_e)

voltage (u_{S1}, u_{L1}), motor current ($i_{S\alpha}$), speed (n) and electromagnetic torque, during the motor starting for output voltage frequency $f_L = 50$ and 75 Hz. Whereas, in Fig. 5.24 the same time waveforms are presented in motor start and reverse states for output voltage frequency $f_L = 50\,\text{Hz}$. All results are obtained for open loop control for sequence duty factor $D_S = 0.8$ [7, 8].

All these results show that the converter voltage (u_{L1}) is greater than the supply voltages (Figs. 5.22a, 5.23a, 5.24a). This results in a faster motor start-up to the nominal speed than direct motor attached to the supply grid (Figs. 5.22c, 5.23c, 5.24c). As is clear from (Figs. 5.22b, 5.23b, 5.24b) during the start-up process and reverse process the stator currents $i_{S\alpha}$ are greater than in steady state. Increased motor power consumption in transient states causes a drop in the MRFC output voltage (u_{L1}). This voltage is fixed after reaching the nominal motor speed $n_{(f_L)}$ for a given frequency of motor supply voltage (f_L). It should be noted that MRFC output voltage is greater than the supply voltage at any time Table 5.2 [7, 8]. In practical solutions of drive systems with a feedback control the output voltage can be maintained at a constant level equal to or greater than the supply voltage. Figures 5.22, 5.23 and 5.24 illustrate that the motor supply voltages have a sinusoidal shape. This property is an advantage of using an MRFC in a drive system.

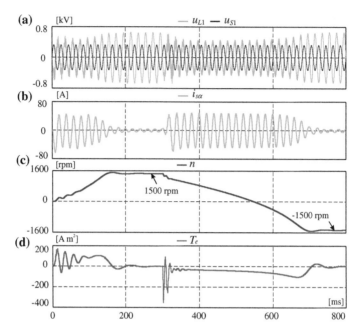

Fig. 5.24 Simulation time waveforms of drive system with MRFC-I-b-b during motor starting and revers, for $f_L = 50\,\text{Hz}$, $D_S = 0.8$, $q = 0.5$: **a** input (u_{S1}) and output (u_{L1}) voltage of converter, **b** motor current ($i_{S\alpha}$), **c** speed (n), **d** electromagnetic torque (T_e)

5.5 Chapter Summary

This chapter presented the basic steady and transient states theoretical results of MRFCs. Due to the large number of converters, only the MRFC-I-b-b topology has been analysed in detail. The theoretical results presented in this chapter, and the exact match between theoretical analysis and numerical simulation that has been achieved, indicate the correctness of the theoretical analysis.

For final comparison of all topologies, the static characteristics of voltage and current gain and input power factor for $f_L = 25\,\text{Hz}$ are demonstrated in Fig. 5.25 [2, 8].

In all the analysed topologies of MRFCs, for modified Venturini control strategy and parameters listed in Table 5.1, it is possible to obtain a voltage gain greater than unity as shown in Fig. 5.25a. The voltage gain, depending on the topology, varies widely in range, and the maximum is over 10 for the MRFC-II-c topology. The level of reactor currents in the selected topology can be analysed based on the current gain characteristic (Fig. 5.25b). Figure 5.25c shows the input power factor of the whole family of MRFCs using a modified Venturini control strategy. In converters with this modulation strategy it is not possible to control the input power factor. The modification of the control strategy is needed for input power factor improvement.

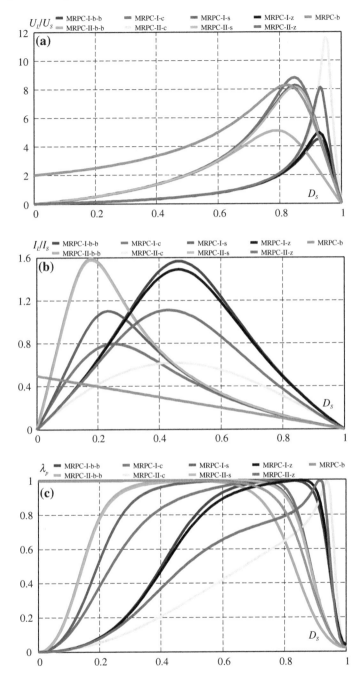

Fig. 5.25 Comparison of static characteristics of: **a** voltage gain K_U, **b** current gain K_I, **c** input power factor λ_P

Table 5.3 Summaries of the maximum voltage gain and input power factor of MRFCs

Topology	Maximum voltage gain			Range D_S, when $\lambda_p > 0.95$		
	$f_L = 25\,\text{Hz}$	$f_L = 50\,\text{Hz}$	$f_L = 75\,\text{Hz}$	$f_L = 25\,\text{Hz}$	$f_L = 50\,\text{Hz}$	$f_L = 75\,\text{Hz}$
MRPC-I-b-b	4,49	3.56	3.09	0.62–0.87	0.65–0.85	0.67–0.85
MRPC-II-b-b	5.10	5.32	5.55	0.28–0.66	0.34–0.69	0.36–0.70
MRPC-I-c	8.78	9.73	10.70	0.56–0.79	0.70–0.81	0.74–0.81
MRPC-II-c	11.68	11.98	12.42	0.93–0.94	0.93–0.94	0.93–0.94
MRPC-I-z	4.94	4.19	3.86	0.67–0.89	0.76–0.87	0.79–0.86
MRPC-II-z	8.24	8.54	8.80	0.89–0.92	0.90–0.92	0.90–0.92
MRPC-I-s	8.26	8.64	9.03	0.38–0.77	0.5–0.79	0.59–0.80
MRPC-II-s	8.14	8.52	8.90	0.30–0.77	0.34–0.78	0.41–0.79
MRPC-b	8.25	8.64	9.04	0.00–0.71	0.00–0.73	0.00–0.75

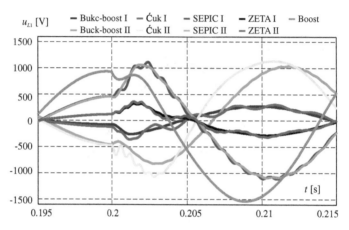

Fig. 5.26 Transient responses of output voltages of MRFCs at step change of the sequence pulse duty factor D_S from 0.5 to 0.7, for $f_L = 25\,\text{Hz}$

Using SVM technique for MRFC control the improvement of their properties is possible [10].

Table 5.3 summarises the maximum voltage gain and input power factor range of slightly less than unity for all MRFCs with modified Venturini control strategy and parameters in Table 5.1 [8]. From this table it can also be seen that there is significant difference between the MRFC topologies.

Moreover, in order to compare the dynamic properties of the whole family of MRFCs, the output voltage time waveforms in transient states for D_S changes from 0.5 to 0.7 for $f_L = 25\,\text{Hz}$ are presented in Fig. 5.26. The time response depends on the topology and the voltage gain for the initial steady state [2, 8].

During the practical implementation and design of the selected MRFC topology the type of load, the maximum voltage gain and the switching frequency of power switches should be considered. In addition, the design of reactors must take into account their current values. To estimate the level of currents, the characteristics of current gain may be useful.

Based on the Obtained Results it can be Concluded that

- MRFCs allow a change of the output voltage frequency and the buck-boost output voltage regulation,
- there are two degrees of control freedom in the output voltage through the coefficient q and sequence pulse duty factor D_S,
- properties of MRFCs depends on the setting frequency of the output voltage,
- properties of MRFCs are sensitive to load changes,
- systems with MRPCs have good dynamic properties,
- in MRFCs with modified Venturini control strategy the input power factor regulation is not possible,
- in design process of MRFCs experimental setup the maximum voltage gain, current level and switching frequency must be taken into account.

References

1. Fedyczak Z (2003) PWM AC voltage transforming circuits (In Polish). Zielona Góra University Press, Zielona Góra
2. Fedyczak Z, Szcześniak P (2012) Matrix-reactance frequency converters using an low frequency transfer matrix modulation method. Electr Power Syst Res 83(1):91–103
3. Fedyczak F, Szcześniak P (2009) Modelling and analysis of matrix-reactance frequency converters using voltage source matrix converter and LF transfer matrix modulation method. Przegląd Elektrotechniczny (Electr Rev) 2:125–130
4. Korotyeyev I, Fedyczak Z (2008) Steady and transient states modelling methods of matrix-reactance frequency converter with buck-boost topology. COMPEL: Int J Comput Math Electr Electron Eng 28(3):626–638
5. Korotyeyev I, Fedyczak Z, Szcześniak P (2008) Steady and transient state analysis of a matrix-reactance frequency converter based on a boost PWM AC matrix-reactance chopper. In: Proceedings of the international school on nonsinusoidal currents and compensation, ISNCC'08, Łagów, Poland (CD-ROM)
6. Szczęsny R (1999) Computer simulation of power electronic systems, (Komputerowa symulacja układów energoelektronicznych) (in Polish). Wydawnictwo Politechniki Gdańskiej, Gdańsk
7. Szcześniak, P (2010) Analiza i badania właściwości układu napędowego z matrycowo reaktancyjnym przemiennikiem częstotliwości o modulacji Venturiniego (in Polish). Przegląd Elektrotechniczny (Electr Rev), 6:155–158
8. Szcześniak P (2009) Analysis and testing matrix-reactance frequency converters. PhD thesis (in Polish), University of Zielona Góra, Zielona Góra
9. Szcześniak P, Fedyczak Z, Klytta M (2008) Modelling and analysis of a matrix-reactance frequency converter based on buck-boost topology by DQ0 transformation. In: Proceedings of power electronics and motion control conference, EPE-PEMC'08, Poznań, Poland, pp 165–172
10. Szcześniak P, Fedyczak Z, Tadra G (2011) Modeling of the matrix-reactance frequency converters using SVM method (in Polish). In: Proceedings of Sterowanie w Energoelektronice i Napędzie Elektrycznym, SENE (2011) Łódź, Poland (CD-ROM)
11. Venturini M, Alesina A (1980) The generalized transformer: a new bi-directional sinusoidal waveform frequency converter with continuously adjustable input power factor. In: Proceedings of IEEE power electronics specialists conference PESC'80, pp 242–252

Chapter 6
Experimental Investigation

6.1 Introduction

Some prototypes of different MC topologies have already been designed and described in the literature [5, 8–11, 14–16].

The MRFC control and commutation algorithms require larger computational capacity. This control will be realised in a DSP and FPGA device. The presented solution uses the Analog Devices floating point processors with Sharc series and Xilinx flexible programmable gate array (FPGA) with the Spartan 3 family. This FPGA provides an interesting alternative for building specific DSP systems.

The modified Venturini modulation strategy has been used in the control and implemented in DSP. Four-step current-based switch commutation is implemented in FPGA devices. The current and voltage measurements are based on LEM transducers. Galvanic separation of power and control stages is provided by a fibre-optic interface. Preliminary tests with an R-passive load have been performed to verify the effectiveness of the solutions adopted. For purpose of comparison some of these results are presented together with ones obtained by means of theoretical investigations of the presented circuits. Experimental test results generally conform to the theoretical ones. Furthermore, the simplified experimental test of a drive system with a 0.4 kW induction cage motor is presented.

6.2 Practical Implementation

The experimental schematic circuit of an MRFC-I-b-b and MRFC-II-b-b feeding a passive load is shown in Figs. 6.1 and 6.2, respectively [7, 13]. The systems comprise a regulated voltage supply, the MRFC, a control circuit and a load impedance.

The reconfigurable MRFC prototype, with specification of the basic components is shown in Fig. 6.3. On the left there is the power board, mounted on a heat sink. On the right there is the power board and computer with the DSP control unit. The MRFC

P. Szcześniak, *Three-Phase AC–AC Power Converters Based on Matrix Converter Topology*, Power Systems, DOI: 10.1007/978-1-4471-4896-8_6,
© Springer-Verlag London 2013

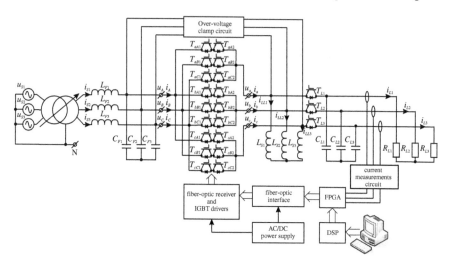

Fig. 6.1 The experimental schematic circuit of an MRFC-I-b-b

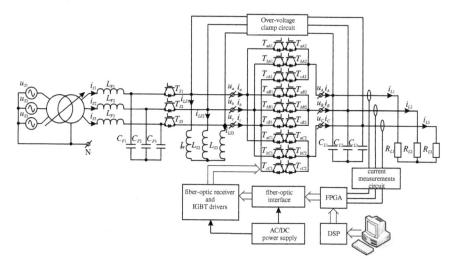

Fig. 6.2 The experimental schematic circuit of an MRFC-II-b-b

requires bidirectional switches with the capability to block the voltage and to conduct the current in both directions. The MRFC prototype has been built with the discrete module IRG4PH50KDPbF. Each module includes one IGBT and one fast recovery diode. The diode provides the reverse blocking capability. The module parameters are as follows: $U_{CES} = 1200\,\text{V}$, $U_{CE(on)} = 2.77\,\text{V}$, $U_{GE} = 15\,\text{V}$, $I_C = 24\,\text{A}$. The bi-directional switches are connected in the common emitter anti-parallel IGBT configuration. Output switches in Fig. 6.1 or input in Fig. 6.2 are constructed as uni-directional switches, and includes one IGBT and one diode. All transistor drivers are

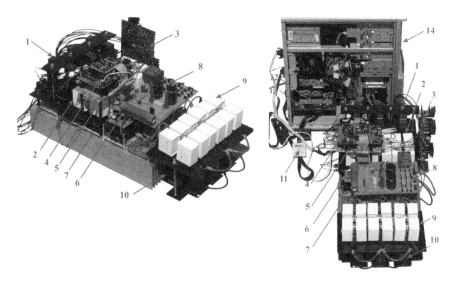

Fig. 6.3 View of the MRCF set-up; *1* source filter inductors; *2* source filter capacitors; *3* AC-DC power supply; *4* FPGA card (ZL9PDL); *5* optical transmitters; *6* output current measurement circuits; *7* optical receivers and gate driver circuits; *8* over-voltage clamp circuit; *9* load capacitors; *10* load inductors; *11* JTAG Emulator (HPPCI-ICE); *12* DSP Card (ALS-G3-2368PCI) *13* AD-DA Card (ALS-G3-ACA1812-1); *14* PC

Table 6.1 MRFC experimental set-up parameters

Parameters	Symbol	Value	
Supply voltage	U_S	MRPC-I-b-b	MRPC-II-b-b
		45 V	20 V
Frequency of supply voltage	f	50 Hz	
Switching frequency	f_S	5 kHz	
Input filter inductance	L_{F1}, L_{F2}, L_{F3}	1.5 mH	
Source inductance	L_{S1}, L_{S2}, L_{S3}	1.5 mH	
Input filter capacitance	C_{F1}, C_{F2}, C_{F3}	10 μF	
Load capacitance	C_{L1}, C_{L2}, C_{L3}	10 μF	
Load resistance	R_{L1}, R_{L2}, R_{L3}	60 Ω	

supplied by local independent insulated power supplies, in order to ensure correct operation. A single common clamp circuit protects all the nine bi-directional switches of the converter against over-voltages from the input and the output sides (Fig. 2.37). The small capacitor of the clamp is designed to store the energy corresponding to the inductive load current with an acceptable over-voltage. The three-phase autotransformer is used to control supply voltages with an amplitude of 60 V. The frequency of the input voltages is 50 Hz. The switching frequency was 5 kHz.

A second order *LC* filter is inserted at the input side of the converter in order to reduce the ripple of the line currents. It consists of three series connected inductors

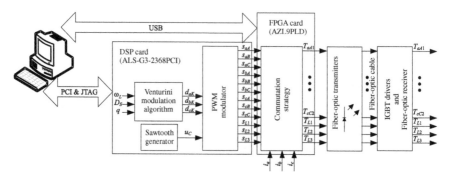

Fig. 6.4 Block scheme of control algorithm

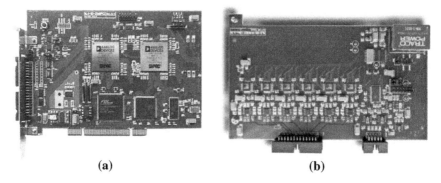

(a)   **(b)**

Fig. 6.5 DSP boards: **a** two-DSP card ALS-G3-2368PCI, **b** A/D-D/A converters card
ALS-G3-ACA1812-1

(1.5 mH/20 A) and three shunt connected capacitors (10 μF/1200 V). The input filter
is generally needed to smooth the input currents and to satisfy the electromagnetic
interference (EMI) requirements. The disadvantages of the input filter presence is
a reduction of input power factor. Additional inductors (L_{S1}, L_{S2}, L_{S3}) and capacitors
(L_{L1}, L_{L2}, L_{L3}) are the same values as in input filters. The output resistance was
set to 60 Ω per phase. The system parameters of the laboratory set-up are listed
in Table 6.1, and all the symbols are given with respect to the schematic circuit in
Figs. 6.1 and 6.2.

The block scheme of control algorithm is shown in Fig. 6.4. The modified
Venturini modulation approach is used to control the matrix converter [6, 7, 13].
The control system of the matrix converter is constituted by two floating-point digi-
tal signal processors (DSP). The adopted DSP is the SHARC ADSP-21368 by Analog
Devices, running at 333 MHz. The DSP is mounted on an evaluation board ALS-
G3-2368PCI manufactured by Alfine P.E.P., which provides the basic interfaces for
the use of the DSP. Figure 6.5a shows the DSP board view. Its basic parameters
are as follows: 2 × 16 PWM output (or 2 × 16 digital I/O), 1 Mb RAM memory and

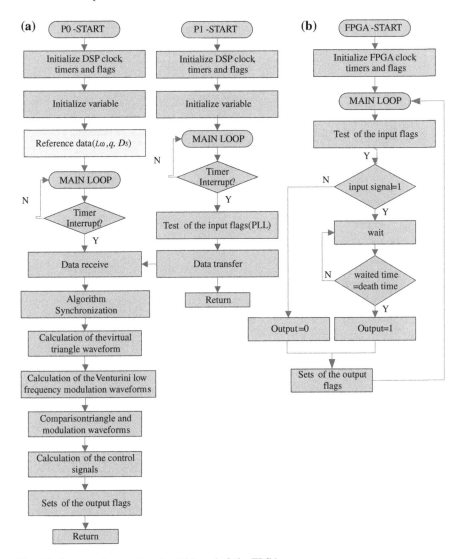

Fig. 6.6 Program flowcharts, **a** for DSP cards, **b** for FPGA

512 kb flash memory. The presented DSP board co-operates with an additional board, the ALS-G3-ACA1812-1 with analog to digital (A/D) and digital to analog D/A converters (Fig. 6.5b), with the following specifications: 6×18-bit A/D Sample Rate -up to 800 kHz; input voltage range—± 2.5 V, one12-bit, 4-channel D/A converter with output voltage range—± 2.5 V and settling time of $-5\mu s$ (the A/D-D/A board will be using an SVM algorithm). A standard PC computer provides the user with an interface to the DSP via conventional PCI computer bus (Peripheral Component Interconnect). *VisualDSP++4.5* software is used for DSP programming.

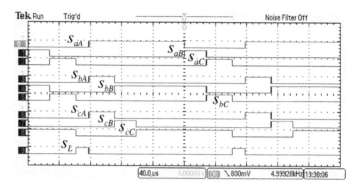

Fig. 6.7 DSP control signals for $D_S = 0.9$, $q = 0.5$ and $f_L = 50$ Hz

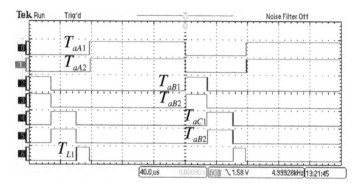

Fig. 6.8 FPGA control signals for $D_S = 0.9$, $q = 0.5$ and $f_L = 50$ Hz

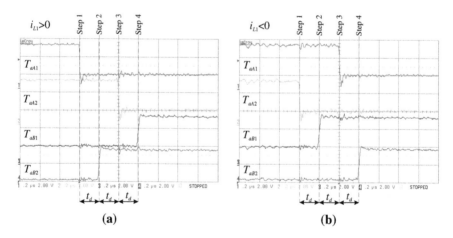

Fig. 6.9 Experimental time waveforms of four-step direction current based commutation strategy: **a** for $i_L > 0$, **b** $i_L < 0$

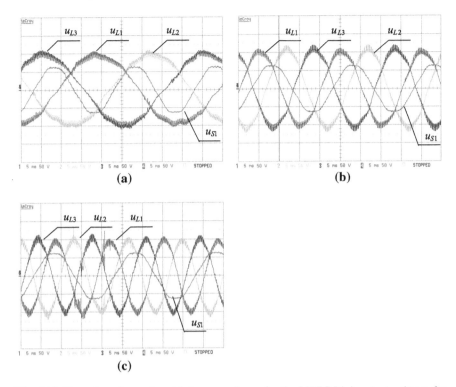

Fig. 6.10 Examples of experimental time waveforms for the MRFC-I-b-b output voltages for $D_S = 0.8$: **a** $f_L = 25\,$Hz, **b** $f_L = 50\,$Hz, **c** $f_L = 75\,$Hz

The calculation of the switch duty-cycle is performed by the presented DSP system. The control algorithm described in the Sect. 3.5 has been programmed in C-language and Assembler. The control algorithm used 10 PWM outputs, set to 5 kHz frequency. The algorithm is initialised by an interrupt routine taken from a PWM module. The interrupt is initiated when the PWM counter is reset. The PWM counter operates in down mode. This mode allows the generation of PWM pulses that are required in Venturini modulation, synchronised at the beginning of the switching sequence [6, 7, 13]. The analog-to-digital converters A/D and D/A are also synchronised with PWM counter. The overall structure of the DSP algorithm is shown in Fig. 6.6a.

Control signals from the microprocessor are transmitted to the FPGA (Field Programmable Gate Array) control board—ZL10PLD. This device is based on an XC3S200 unit belonging to the Spartan 3 family of Xilinx. Its basic parameters are as follows: 216 kb SRAM memory, 1 Mb flash memory and 61 digital I/O. The FPGA is responsible for processing the signals produced by the DSP and generating 21 IGBT gate drive pulses according to certain commutation methods. The four-step current direction-based commutation is implemented for this MRFC prototype. The LEM sensors are used to obtain the signs of output currents. The output current direction

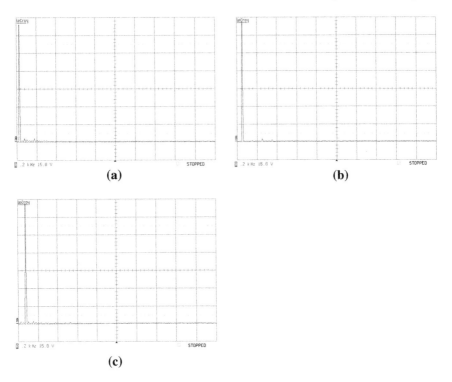

Fig. 6.11 Spectrum of the MRFC-I-b-b output voltages for $D_S = 0.8$: **a** $f_L = 25\,\text{Hz}$, **b** $f_L = 50\,\text{Hz}$, **c** $f_L = 75\,\text{Hz}$

signal is assumed as "1" when current flows into the load and "0" for the reverse direction. The commutation time is about $0.2\,\mu\text{s}$. The overall structure of the DSP algorithm is shown in Fig. 6.6b. Control signals from the FPGA are transmitted to the IGBT driver devices through a fibre-optic interface. In general, the DSP provides fast calculations for modulation algorithms, while the FPGA is used for advanced timing functions.

Exemplary experimental control signals from DSP and FPGA devices are presented in Figs. 6.7 and 6.8 respectively. Figure 6.7 shows ten control signals of all MRFC switches, whereas Fig. 6.8 shows seven control signals for one-phase transistors, taking into account the four-step commutation algorithm. The four-step switch commutation for two bi-directional switches in experimental set-up are presented in Fig. 6.9. The one-step delay time is set to $0.2\,\mu\text{s}$. The sequence of commutation process lasts $0.6\,\mu\text{s}$. The commutation algorithm is applied in matrix connected switch sets, whereas in switch sequence the dead time is set between control signals of additional switches (S_{L1}, S_{L2}, S_{L3} for MRFC-I-b-b or S_{S1}, S_{S2}, S_{S3} for MRFC-II-b-b) and matrix connected switch sets. The IGBT drivers operate with the level of control signal voltage $+15 -5\,\text{V}$, whereas DSP and FPGA operate with $+5\,\text{V}$ signal level.

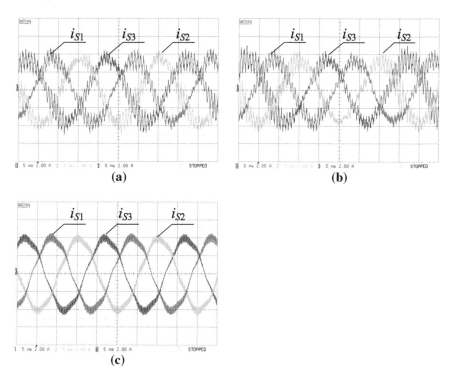

Fig. 6.12 Examples of experimental time waveforms for the MRFC-I-b-b input currents for: **a** $f_L = 25\,\text{Hz}$, **b** $f_L = 50\,\text{Hz}$, **c** $f_L = 75\,\text{Hz}$

6.3 Experimental Results

Preliminary tests with an R-passive load have been performed to verify the properties of MRFCs. Two topologies of MRFC-I-b-b and MRFC-II-b-b have been tested. Experimental circuit schemes are presented in Figs. 6.1 and 6.2.

Figure 6.10 shows the MRFC-I-b-b phase load voltage for a voltage transfer ratio $q = 0.5$, sequence pulse duty factor $D_S = 0.8$ and different output voltage frequencies of $f_L = 25, 50, 75\,\text{Hz}$. The spectra of the presented output voltage are shown in Fig. 6.11. As will be noticed, on the time waveforms and spectra for output voltages there occur low frequency distortions in the sinusoidal shape. They result from the distortion of source voltage as shown in Fig. 6.10. The system can be connected directly to the public power grid in the laboratory. The voltages do not have a sinusoidal shape in this power grid. There are low frequency distortions (3rd and 5th order harmonics). The accuracy of all power circuit components is also very important for good performance of output voltages. Some power circuit components such as inductors, are made with low-precision. The implemented control strategy has good performance only in balanced conditions and with sinusoidal source voltages. A control strategy which reduces unbalanced input and output conditions is needed

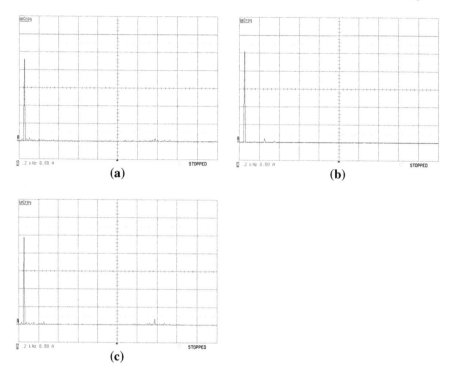

Fig. 6.13 Spectrum of the MRFC-I-b-b input currents for $D_S = 0.8$: **a** $f_L = 25\,$Hz, **b** $f_L = 50\,$Hz, **c** $f_L = 75\,$Hz

to obtain better performances [1–4]. Also, high frequency distortions are visible in output voltage time waveforms. Their cause is the switching frequency of power transistors.

The time waveforms shown in Fig. 6.10 confirm that by means of the discussed MRFC circuit (Fig. 6.1) frequency conversion and buck-boost load voltage changes are possible. The obtained output voltages are much larger than the input one (Fig. 6.10).

Figure 6.11 illustrates the source current obtained in the experimental laboratory model. In Fig. 6.12 the harmonic spectrum for the input line current is compared for three output setting frequencies. The wide oscillations of the input line current are due to a resonance phenomenon which occurs between the impedance of the MRFC passive RLC components. The best results are obtained for $f_L = 50\,$Hz (Fig. 6.11b). For the other frequencies, the input currents have higher frequency oscillations (Fig. 6.11a, c), with frequencies near the input filter cutoff frequency $f = 1400\,$Hz. Figures 6.12a, c show that the current harmonic components with higher amplitude are centred around the input filter cutoff frequency.

Similar voltage and current time waveforms and harmonic spectrums are presented for the second experimental model of MRFC-II-b-b whose experimental

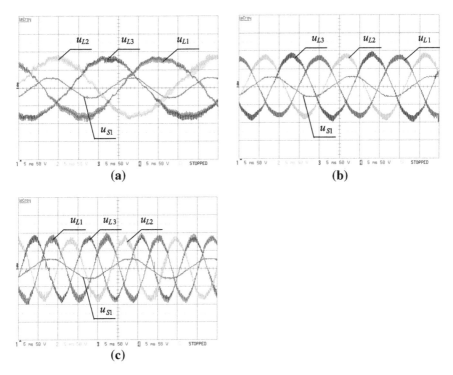

Fig. 6.14 Examples of experimental time waveforms for the MRFC-II-b-b output voltages for $D_S = 0.7$: **a** $f_L = 25\,\text{Hz}$, **b** $f_L = 50\,\text{Hz}$ **c** $f_L = 75\,\text{Hz}$

circuit scheme is indicated in Fig. 6.2. The output phase voltage and source current for a voltage transfer ratio $q = 0.5$, sequence pulse duty factor $D_S = 0.7$ and different output voltage frequencies $f_L = 25, 50, 75\,\text{Hz}$ are presented in Figs. 6.13 and 6.15, respectively, whereas the harmonics spectrums are shown in Figs. 6.14 and 6.16. Also, the obtained time waveforms shown in Fig. 6.13 confirm that by means of the discussed MRFC circuit (Fig. 6.2) frequency conversion and buck-boost load voltage changes are possible (Fig. 6.13). The voltage gain in a system with MRFC-II-b-b is greater than with MRFC-I-b-b. Also, the voltage and current performance are better than in MRFC-I-b-b. The input current time waveforms have a lower level of distortion Fig. 6.15. The low frequency distortion of source voltages also has an influence on the obtained results of voltages and currents (Figs. 6.14, 6.16, 6.17).

In order to test the behaviour of the converter after a sudden sequence pulse duty factor D_S variation the reference D_S has been changed from 0.8 to 0.5 and from 0.5 to 0.8. Figure 6.18 shows the MRFC-I-b-b load voltage during the transient. The transient period is very small. The load voltage variation is almost immediate. As can be seen, the output voltage remains nearly sinusoidal. Figure 6.19 shows the output voltage waveforms during operation in output frequency f_L variation. It shows the

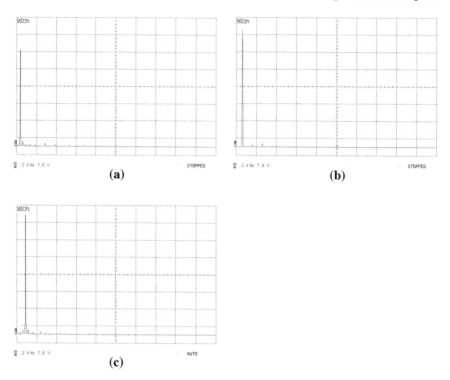

Fig. 6.15 Spectrum of the MRFC-II-b-b output voltages for $D_S = 0.7$: **a** $f_L = 25\,Hz$, **b** $f_L = 50\,Hz$, **c** $f_L = 75\,Hz$

response to an output frequency rise from 25 to 75 Hz and dips from 75 to 25 Hz. The circuit response is small.

In Figs. 6.20 and 6.21 the comparison of the selected calculation and experimental test results in the form of the voltage gain, current gain and input power factor static characteristics for both MRFCs based on buck-boost topology is illustrated. In Figs. 6.20d and 6.21d the results of the experimental investigations of the efficiency coefficient for both MRFCs are shown. As is visible from these figures the obtained results are not favourable yet. There is a low value of efficiency caused by the inaccuracy in the converter implementation and, therefore, for safety reasons there is a reduced voltage in the constructed prototypes. In the general case the results of the experimental investigation confirm the results of theoretical studies. Differences between analytic and experimental results are caused by higher harmonics being taken into account for the experimental work and low efficiency of the experimental set-up. In theoretical analysis the parasitic capacitive inductances and connection resistances are not taken into consideration.

A prototype MRFC-I-b-b has been tested on a three-phase 0.4 kW, 2-pole, 50 Hz cage induction motor [12, 13]. A view of the laboratory set-up is shown in Fig. 6.22. To illustrate the motor start-up process the time waveforms of speed (n), motor

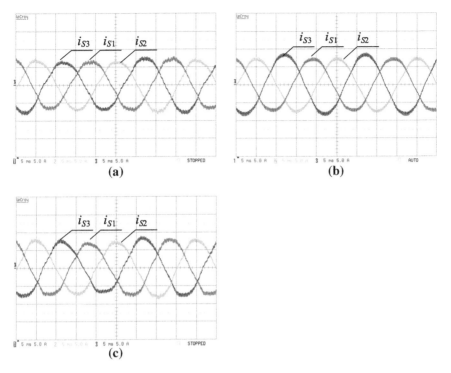

Fig. 6.16 Examples of experimental time waveforms for the MRFC-II-b-b input currents for $D_S = 0.7$: **a** $f_L = 25\,\text{Hz}$, **b** $f_L = 50\,\text{Hz}$, **c** $f_L = 75\,\text{Hz}$

current (i_{S1}) and converter output voltage (u_{L1}) for two frequencies $f_L = 25$ and 50 Hz are shown in Fig. 6.23. All the results are obtained for open-loop control for sequence duty factor $D_S = 0.8$.

The converter output voltage u_{L1} is greater than the supply voltages as shown in the voltage time waveforms zoom presented in Fig. 6.24. As can be seen in Fig. 6.23 during the start-up process the stator currents i_{S1} are greater than when in steady state. Increased motor power consumption in transient states causes a drop in the MRFC output voltage (u_{L1}). This voltage is fixed after reaching the nominal motor speed $n_{(f_L)}$ for a given frequency of motor supply voltage. It should be noted that MRFC output voltage is greater than the supply voltage at any time, similar to that shown in simulation results (Figs. 5.21, 5.22, 5.23).

6.4 Chapter Summary

The design and implementation of two MRFCs which are based on an MRC with buck-boost topology have been presented in this chapter. Furthermore, the experimental test results for passive R load and induction cage motor have been shown.

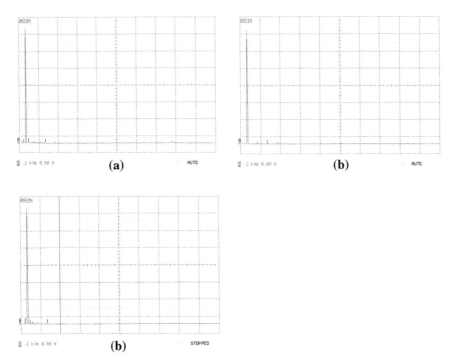

Fig. 6.17 Spectrum of the MRFC-II-b-b input currents for $D_S = 0.7$: **a** $f_L = 25\,\text{Hz}$, **b** $f_L = 50\,\text{Hz}$, **c** $f_L = 75\,\text{Hz}$

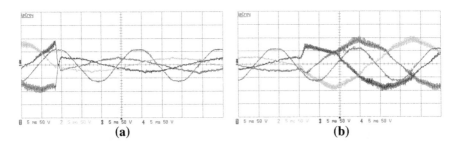

Fig. 6.18 Experimental output voltage time waveforms of MRPC-I-b-b at change of sequence pulse duty factor D_S for $fL = 25\,\text{Hz}$, $q = 0.5$; **a** from 0.8 to 0.5 Hz, **b** from 0.5 to 0.8 Hz

Overall, the experimental test results of the discussed MRFCs confirm the theoretical ones presented in Chap. 5.

This chapter has been mainly focused on the technical solutions adopted to obtain good performance of the converters. Some practical implementation issues have been presented. The MRFC prototypes work, for safety reasons, with a reduced supply voltage. Future research dealing with the discussed MRFCs will focus on design and implementation of devices using the full range of the power supply. The experimental laboratory models that have been built serve primarily to verify the MRFC concept.

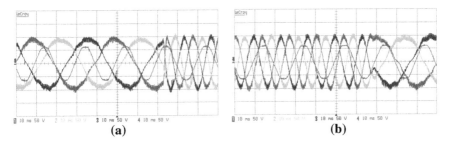

Fig. 6.19 Experimental output voltage time waveforms of MRPC-I-b-b at change of output frequency f_L for $D_S = 0.8$, $q = 0.5$: **a** from 25 to 75 Hz, **b** from 75 to 25 Hz

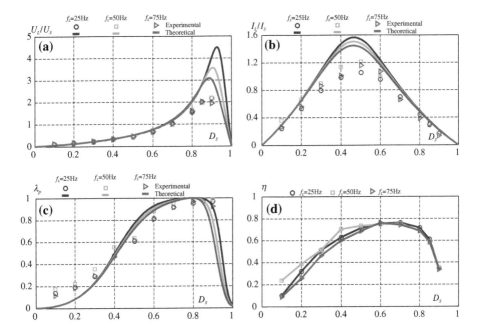

Fig. 6.20 Experimental and theoretical MRFC-I-b-b static characteristics of: **a** voltage gain, **b** current gain, **c** input power factor, **d** efficiency coefficient

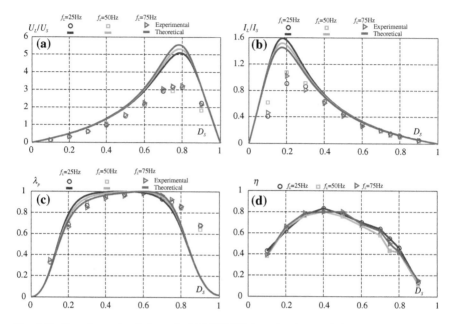

Fig. 6.21 Experimental and theoretical MRFC-II-b-b static characteristics of: **a** voltage gain, **b** current gain, **c** input power factor, **d** efficiency coefficient

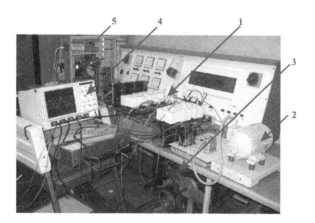

Fig. 6.22 Laboratory model of drive system with MRFC-I-b-b and 0.4 kW cage motor; *1* MRFC-I-b-b, *2* 0.4 kW cage motor, *3* autotransformer, *4* oscilloscope, *5* PC computer with control circuit

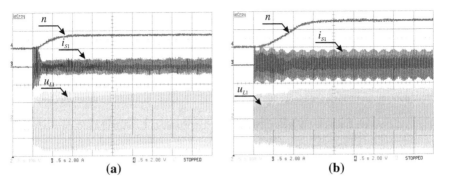

Fig. 6.23 Experimental motor signals during start-up for: **a** $f_L = 25\,\text{Hz}$, **b** $f_L = 50\,\text{Hz}$

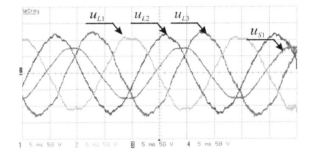

Fig. 6.24 Motor voltages for $f_L = 50\,\text{Hz}$

References

1. Casadei D (2005) Tutorial on matrix converters. In: Proceedings of power electronics and intelligent control for energy conservation conference, PELINCEC'05, Warsaw, Poland
2. Casadei D, Grandi G, Serra G, Tanti A (1993) Space vector control of matrix converters with unity input power factor and sinusoidal input/output waveforms. In: Proceedings of European conference on power electronics and applications, EPE'93, vol 7, Brighton, UK, pp 170–175
3. Casadei D, Serra G, Tani A (1998) Reduction of the input current harmonic content in matrix converters under input/output unbalance. IEEE Trans Ind Electron 45(3):401–411
4. Casadei D, Serra G, Tani A, Nielsen P (1995) Performance of SVM controlled matrix converter with input and output unbalanced conditions. In: Proceedings of European conference on power electronics and applications, EPE'95, vol 2, Seville, Spain, pp 628–633
5. Casadei D, Serra G, Tani A, Zarri L (2005) Experimental behavior of a matrix converter prototype based on new power modules. Automatika (J Control Meas Electron Comput Commun) 46(1–2):83–91
6. Fedyczak Z, Szcześniak P (2012) Matrix-reactance frequency converters using an low frequency transfer matrix modulation method. Electr Power Syst Res 83(1):91–103
7. Fedyczak Z, Szcześniak P, Kaniweski J, Tadra G (2009) Implementation of three-phase frequency converters based on PWM AC matrix-reactance chopper with buck-boost topology. In: Proceedings of European conference on power electronics and applications, EPE'09, Barcelona, Spain, pp P1–P10 (CD-ROM)

8. Jussila M, Salo M, Tuusa H (2003) Realization of a three-phase indirect matrix converter with an indirect vector modulation method. In: Proceedings of power electronics specialist conference, PESC'03, vol 2, Acapulco, Meksyk, pp 689–694

9. Klumpner C, Nielsen P, Boldea I, Blaabjerg F (2002) A new matrix converter motor (MCM) for industry applications. IEEE Trans Ind Electron 49(2):325–335

10. Lee MY, Klumpner C, Wheeler PW (2008) Experimental evaluation of the indirect three-level sparse matrix converter. In: Proceedings of IET international conference on power electronics, machines and drives, PEMD'08, York, UK, pp 50–54

11. Podlesak TF, Katsis D, Wheeler PW, Clare J, Empringham L, Bland M (2005) A 150 kVA vector controlled matrix converter induction motor drive. IEEE Trans Ind Appl 41(3):841–847

12. Szcześniak P (2010) Analiza i badania właściwości układu napędowego z matrycowo reaktancyjnym przemiennikiem częstotliwości o modulacji Venturiniego (in Polish). Przegląd Elektrotechniczny (Electr Rev) 6:155–158

13. Szcześniak P (2009) Analysis and testing matrix-reactance frequency converters. PhD thesis (in Polish), University of Zielona Góra, Zielona Góra

14. Wheeler PW, Clare JC, Apap M, Empringham L, Bradley KJ, Pickering S, Lampard DA (2005) Fully integrated 30kW motor drive using matrix converter technology. In: Proceedings of European conference on power electronics and applications, EPE'05, Dresden, pp 2390–2395

15. Wijekoon T, Klumper C, Zanchetta P, Wheeler PW (2008) Implementation of a hybrid AC-AC direct power converter with unity voltage transfer. IEEE Trans Power Electron 23(4): 1918–1926

16. Yamamoto E, Hara H, Kang JK, Krug HP (2011) Development of MCs for industrial applications. IEEE Ind Electron Mag 5:4–12

Chapter 7
Summary of Book

This monograph concerns frequency converters without DC energy storage elements. Such converters are an alternative solution to commonly used converters in industrial application with either voltage or current DC energy storage elements. Improvements in power semi-conductor switches over the last few years have resulted in the development of many AC–AC converters structures without DC electric energy storage elements. Therefore, the first part of this book is dedicated to a general review of such converters. Several frequency converter topologies have been presented. Special attention has been given to the matrix converter topology, which is the most widely known. From the structure of the matrix converter, the majority of topologies of frequency converters without DC energy storage elements have been obtained. The matrix converter switch configuration, switch commutation, protection issue and modulation strategies have been presented in detail.

The second part of this book concerns a new type of converter, which is a major area of author's research interest. There has been presented a description of the family of matrix-reactance frequency converters. With these converters, it is possible to control the output voltage with buck-boost mode and frequency change. These properties are the advantages of frequency converters without a DC storage element. In most of such converters, the fact that the output voltage is less than the supply voltage has already been noted in the second chapter.

Some analysis of the results of matrix-reactance frequency converter properties has been presented. This analysis is based on the results of the author's own simulation investigations, which constitute a point of departure for theoretical deliberations and experimental investigations. The monograph includes the following within the scope of its theoretical analysis:

- a description of the modelling procedure based on the average state-space method;
- a construct of the average state-space models for the whole family of MRFCs for modified Venturini control strategy;
- the generalised expressions describing the average state-space models in stationary conditions when using the two-frequency based dq transformation;

P. Szcześniak, *Three-Phase AC–AC Power Converters Based on Matrix Converter Topology*, Power Systems, DOI: 10.1007/978-1-4471-4896-8_7, © Springer-Verlag London 2013

- the generalised expressions describing the solution of the stationary average state-space equations in steady and transient states;
- a determination of the static characteristics of the basic properties of the converters;
- a determination of the static and transient time waveforms;
- comparison of MRFC properties.

Within the scope of simulation research:

- a comparison of the theoretical and simulation results and determination of the usefulness and accuracy of the obtained mathematical models;
- simulation research of MRFC-I-b-b in drive system with induction cage motor.

Within the scope of experimental research:

- experimental verification of the matrix-reactance frequency converter concept for passive load;
- comparison of theoretical and experimental results;
- experimental verification of MRFCs usefulness in a drive system with induction cage motor.

All the obtained results have confirmed that matrix-reactance frequency converters have some interesting properties, such as buck-boost output voltage regulation, and that they can be used in practical implementation in industry.

The literature lists an increasing number of potential applications for frequency converters without DC energy storage elements, as:

- drive systems [13, 17, 19, 22, 24, 25, 27, 28];
- power interfaces of distributed energy sources [1, 3, 6, 7, 12, 18, 20, 23, 28];
- flexible alternating current transmission system applications [6, 9–11, 15, 16];
- plasma control power supplies [14, 21];
- aircraft and deep-sea or space system applications [2, 4, 5, 8, 26].

The analysis of matrix-reactance frequency converters for potential applications should be the subject of ongoing further research. Moreover, such further studies will be focused on:

- implementation of the modified space vector modulation in order to improve the discussed MRFCs properties;
- implementation of other control strategies;
- implementation of advanced switch commutation methods and protection methods;
- implementation of new power switch devices;
- comparative research of other frequency converters, with and without DC storage;
- experimental application on a full range of power supplies;
- efficiency coefficient improvement in experimental models.

References

1. Agarwal V, Aggarwal RK, Patidar P, Patki C (2010) A Novel scheme for rapid tracking of maximum power point in wind energy generation systems. IEEE Trans Energy Convers 25(1):228–236
2. Arevalo SL, Zanchetta P, Wheeler PW, Trentin A, Empringham L (2010) Control and implementation of a matrix-converter-based AC ground power-supply unit for aircraft servicing. IEEE Trans Ind Electron 57(6):2076–2084
3. Barakati SM (2008) Applications of matrix converters for wind turbine systems, VDM Verlag, Berlin
4. Bhangu BS, Snary P, Bingham CM, Stone DA (2005) Sensorless control of deep-sea ROVs PMSMs excited by matrix converter. In: Proceedings of the European conference on power electronics and applications, EPE 2005, Dresden, Germany (CD-ROM)
5. Bucknall RWG, Ciaramella KM (2010) On the conceptual design and performance of a matrix converter for marine electric propulsion. IEEE Trans Power Electron 25(6):1497–1508
6. Cardenas R, Pena R, Clare J, Wheeler P (2011) Analytical and experimental evaluation of a WECS based on a bage induction generator fed by a matrix converter. IEEE Trans Energy Convers 26(1):204–215
7. Chakraborty S, Kramer B, Kroposki B (2009) A review of power electronics interfaces for distributed energy systems towards achieving low-cost modular design. Renew Sustain Energy Rev 13:2323–2335
8. Empringham L, de Lillo L, Khwan-On S, Brunson C, Wheeler PW, Clare JC (2011) Enabling technologies for matrix converters in aerospace applications. In: Proceedings of international conference-workshop compatibility and power electronics, CPE'2011, Tallinn, Estonia, pp 451–456
9. Fedyczak Z, Tadra G, Szczesniak P (2010) Three-phase AC systems interfaced by current source matrix converter with space vector modulation. In: International school on nonsinusoidal currents and compensation, ISNCC'2010, Łagów, Poland, pp 107–112
10. Itoh J-I, Tamada S (2007) A novel engine generator system with active filter and UPS functions using a matrix converter. In: Proceedings of European conference on power electronics and applications, EPE'2007, Aalborg, Denmark, pp 1–10 (CD-ROM)
11. Jahangiri A, Radan A, Haghshenas M (2010) Synchronous control of indirect matrix converter for three-phase power conditioner. Electr Power Syst Res 80(7):857–868
12. Keyhani A, Marwali MN, Dai M (2009) Integration of green and renewable energy in electric power systems. Wiley, New York
13. Lee KB, Blaabjerg F (2008) Simple power control for sensorless induction motor drives fed by a matrix converter. IEEE Trans Energy Convers 23(3):781–788
14. Liu X, Nakamura K, Jiang Y, Yoshisue T, Mitarai O, Hasegawa M, Tokunaga K, Zushi H, Hanada K, Fujisawa A, Idei H, Kawasaki S, Nakashima H, Higashijima A, Araki K (2011) Study of matrix converter as a current-controlled power supply in QUEST tokamak. Plasma Fusion Res 6
15. Monteiro J, Silva JF, Pinto SF, Palma J (2009) Direct power control of matrix converter based unified power flow controllers. In: Proceedings of IEEE industrial electronics conference, IECON'09, Porto, Portugal, pp 1525–1530
16. Monteiro J, Silva JF, Pinto SF, Palma J (2011) Matrix converter-based unified power-flow controllers: advanced direct power control method. IEEE Trans Power Deliv 26(1):420–430
17. Ortega C, Arias A, Caruana C, Balcells J, Asher GM (2010) Improved waveform quality in the direct torque control of matrix-converter-fed PMSM drives. IEEE Trans Ind Electron 57(6):2101–2110
18. Pena R, Cardenas R, Reyes E, Clare J, Wheeler P (2011) Control of a doubly fed induction generator via an indirect matrix converter with changing DC voltage. IEEE Trans Ind Electron 58(10):4664–4674
19. Podlesak TF, Katsis D, Wheeler PW, Clare J, Empringham L, Bland M (2005) A 150 kVA vector controlled matrix converter induction motor drive. IEEE Trans Ind Appl 41(3):841–847

20. Savaghebi M, Dehghani MT, Hooshyar H, Jalilian A (2010) Enhancement of microturbine-generator output voltage quality through application of matrix converter interface. In: Proceedings of international symposium on power electronics electrical drives automation and motion, SPEEDAM'2010, Pisa, Italy, pp 1823–1826

21. Shimada K, Itoh J-I, Matsukawa M, Kurihara K (2007) A control method of matrix converter for plasma control coil power supply. Fusion Eng Des 82:1513–1518

22. Simon O, Mahlein J, Muenzer MN, Bruckmarm M (2002) Modern solutions for industrial matrix-converter applications. IEEE Trans Ind Electron 2:401–406

23. Teodorescu R, Liserre M, Rodriguez P (2011) Grid converters for photovoltaic and wind power systems. Wiley-IEEE, New York

24. Vargas R, Ammann U, Hudoffsky B, Rodriguez J, Wheeler P (2010) Predictive torque control of an induction machine fed by a matrix converter with reactive input power control. IEEE Trans Power Electron 25(6):1426–1438

25. Wheeler PW, Clare JC, Apap M, Empringham L, Bradley KJ, Pickering S, Lampard DA (2005) Fully integrated 30 kW motor drive using matrix converter technology. In: Proceedings of European conference on power electronics and applications, EPE'05, Dresden, pp 2390–2395

26. Wheeler PW, Empringham L, Apap M, de Lilo L, Clare JC, Bradley K, Whitley C (2003) A matrix converter motor drive for an aircraft actuation system. In: Proceedings of the European conference on power electronics and applications, EPE'03, Toulouse, France (CD-ROM)

27. Xiao D, Rahman FM (2010) Implementation of sensorless direct torque control using matrix converter fed Interior permanent magnet synchronous motor. In: International power electronics conference, IPEC'2010, Sapporo, Japan, pp 3065–3071

28. Yamamoto E, Hara H, Kang JK, Krug HP (2011) Development of MCs for industrial applications. IEEE Ind Electron Mag 5:4–12

Index

A

Averaged state space equations
  general form, 108
  for matrix-reactance frequency converters
    nonstationary form, 108–110
    solution, 112
    stationary form, 110, 112
Averaged state space models, 108, 124

B

Back-to-back converter, 20
Bi-directional switches, 25–28

C

Commutation
  dead-time, 30
  four-step based on current direction, 31
  four-step based on voltage sign, 35
  general rules, 29
  METZI, 36–38
  overlap method, 30
  switches with RB-IGBT, 39
  three steps, 38
  two-step based on current direction, 32
  with auxiliary resonant components, 42–43
  with intelligent gate driver, 33–35
  with resonant switches cell, 40
Current source inverter, 19–22
Current source matrix converter, 23, 61–64, 88

D

DC energy storage element, 17, 19–21, 74, 76
Drive system, 140–145, 163
DSP, 154, 158

E

Electromagnetic torque, 143
Experimental model, 151, 163

F

FPGA, 157, 158
Frequency converter
  based on matrix-reactance choppers, 72
  hybrid, 74–77
  with DC energy storage element, 19–22
  without DC energy storage element, 22–74

H

Hybrid converter
  direct MC and H-bridge inverter, 74
  indirect MC and H-bridge inverter, 74
  indirect MC with an auxiliary
    DC source, 76

I

IGBT, 26–28, 151
IGBT power module, 27–28

P. Szcześniak, *Three-Phase AC–AC Power Converters Based on Matrix
Converter Topology,* Power Systems, DOI: 10.1007/978-1-4471-4896-8,
© Springer-Verlag London 2013

I (*cont.*)
Indirect matrix converter, 65
Inverting link matrix converter, 70

L
Load-matching condition, 134
Low-frequency transfer matrix, 48–50, 100

M
Matrix converter, 23–61
Matrix-reactance choppers, 89
Matrix-reactance frequency converters
    concept, 87, 88, 90, 92, 94, 96, 98,
        100, 102, 104
    control strategies, 100–103
    MRFC-b, 91, 135
    MRFC-I-b-b, 92, 96–98, 113–118,
        128–135, 137, 151, 158–162
    MRFC-I-c, 93, 129
    MRFC-I-s, 95, 135
    MRFC-I-z, 94, 129
    MRFC-II-b, 135, 163
    MRFC-II-b-b, 92, 99–100, 129,
        151, 158–161
    MRFC-II-c, 93, 129
    MRFC-II-s, 95, 135
    MRFC-II-z, 94, 135
    topologies generation, 87–94
Modular matrix converter, 74
Modulation matrix, 48–50, 55, 100
Modulation techniques
    indirect, 54
    scalar, 52–54
    space vector, 55–61
    venturini
        classical, 49
        improved, 50
Multilevel indirect matrix converter, 71
Multilevel matrix converter, 64

O
One-cycle switched circuit model, 90

P
Protection
    with 6 diode protected clamp circuit, 45
    with 12 diode protected clamp circuit, 44
    with varistor, 45

R
RB-IGBT, 26–28

S
Space vector modulation, 21–22
Sparse matrix converter, 66
Switch configuration, 21, 56–58, 61, 66

T
Transformation dq, 110
Transient state analysis, 135, 137, 161
Two frequency dq transformation, 110, 116,
    118
Two-level indirect frequency
    convertert, 19–20

U
Ultra sparse matrix converter, 66

V
Vector, 21, 56–61
Very sparse matrix converter, 69
Voltage source inverter, 19–22
Voltage source matrix converter, 23, 88